D1084762

Networks and Grids
Technology and Theory

Information Technology: Transmission, Processing, and Storage

Series Editors: Robert Gallager
                    Massachusetts Institute of Technology
                    Cambridge, Massachusetts

                    Jack Keil Wolf
                    University of California at San Diego
                    La Jolla, California

Networks and Grids: Technology and Theory
Thomas G. Robertazzi

CDMA Radio with Repeaters
Joseph Shapira and Shmuel Miller

Digital Satellite Communications
Giovanni Corazza, ed.

Immersive Audio Signal Processing
Sunil Bharitkar and Chris Kyriakakis

Digital Signal Processing for Measurement Systems: Theory and Applications
Gabriele D'Antona and Alessandro Ferrero

Coding for Wireless Channels
Ezio Biglieri

Wireless Networks: Multiuser Detection in Cross-Layer Design
Christina Comaniciu, Narayan B. Mandayam and H. Vincent Poor

The Multimedia Internet
Stephen Weinstein

MIMO Signals and Systems
Horst J. Bessai

Multi-Carrier Digital Communications:
Theory and Applications of OFDM, 2nd Ed
Ahmad R.S. Bahai, Burton R. Saltzberg and Mustafa Ergen

Performance Analysis and Modeling of Digital Transmission Systems
William Turin

Wireless Communications Systems and Networks
Mohsen Guizani

Interference Avoidance Methods for Wireless Systems
Dimitrie C. Popescu and Christopher Rose

Stochastic Image Processing
Chee Sun Won and Robert M. Gray

Coded Modulation Systems
John B. Anderson and Arne Svensson

Communication System Design Using DSP Algorithms:
With Laboratory Experiments for the TMS320C6701 and TMS320C6711
Steven A. Tretter

A First Course in Information Theory
Raymond W. Yeung

Networks and Grids

Technology and Theory

Thomas G. Robertazzi

Stony Brook University

Stony Brook, NY, USA

 Springer

Thomas G. Robertazzi
Department of Electrical and Computer Engineering
Stony Brook University
Stony Brook, NY 11794-2350
tom@ece.sunysb.edu

Library of Congress Control Number: 2007924088

ISBN-13: 978-0-387-36758-3   e-ISBN-13: 978-0-387-68235-8

Printed on acid-free paper.

Mathematics Subject Classification (2000): 90B18, 68M12, 68M20, 60K25, 90B22, 90B35, 94BXX, 68Q60

9 8 7 6 5 4 3 2 1

springer.com   (KeS/EB)

To My Wonderful Parents
Frank and Marie

Preface

Computer networks have assumed an increasing amount of importance in today's world. Grid technology has great potential, although it is more recent and quite a technical challenge. Networking courses in particular are popular with students who, even in college, sense the field's importance and excitement.

The purpose of this book is to provide an undergraduate/first-year graduate text suitable for a computer networks and grid course with a mathematical flavor. Although many books on networking exist, most have little mathematical content.

To some extent this book is based on undergraduate and graduate computer networks courses I have taught since 1983. The book starts with an introductory networking technology chapter. Chapter 2 covers fundamental stochastic (i.e., random) models for networking. Chapter 3 provides an introduction to queueing theory, a widely used tool for modeling and predicting the performance of networked systems. In chapter 4, some fundamental deterministic algorithms for networking are studied. These algorithms include shortest path routing, protocol verification, and error checking codes. Finally, chapter 5 provides an extensive tutorial on divisible load scheduling theory, a relatively new performance evaluation methodology that has applications to grid computing.

At the undergraduate level, I teach the quantitative material (say, chapters 2 and 4 and parts of 3) first while students are fresh and more receptive and save the qualitative technology description for the second half of the course. At the graduate level, one can focus more on the quantitative material (including chapter 5), while allowing students more independence in learning the technology.

In terms of acknowledgments, some edited text and/or original or redrawn figures from ACTA Press, IEEE Publications, and Foundations of Computing and Decision Sciences have been incorporated into this book. I would like to acknowledge the assistance of Profs. Scott Smolka and Wendy Tang in reading certain sections of the manuscript. This work has benefited from its use by my students at Stony Brook. My thanks go to Carlos Gamboa, Jui Tsun Hung, Shevata Kaul, Mequanint Moges, and Harpreet Pruthi for creating some of the figures. Parts of this manuscript were typed by Sandy Pike and Rachel Robertazzi to whom I am thankful for a great job. The assistance of Carlos Gamboa in helping format the manuscript is very much appreciated. In general, my work at Stony Brook has been made easier by the secretarial skills of Judy Eimer, Carolyn Huggins, Debbie Kloppenburg, and the late Maria Krause. This book has benefitted from the excellent editorial and production efforts of Vaishali Damle, Julie Park, and Ann Kostant of Springer. I am very much appreciative of the bemused acceptance by my family, Marsha, Rachel, and Deanna, of my writing efforts. Finally I dedicate this book to my wonderful parents, Frank and Marie.

Stony Brook, NY   *Thomas Robertazzi*
                                                                   January 2007

Contents

Preface vii

1   A Tour through Networking and Grids   1
   1.1   Achieving Connectivity   2
       1.1.1   Coaxial Cable   2
       1.1.2   Twisted Pair Wiring   3
       1.1.3   Fiber Optics   3
       1.1.4   Microwave Line of Sight   4
       1.1.5   Satellites   4
       1.1.6   Cellular Systems   5
       1.1.7   Ad Hoc Networks   7
       1.1.8   Wireless Sensor Networks   8
   1.2   Multiplexing   9
       1.2.1   Frequency Division Multiplexing (FDM)   9
       1.2.2   Time Division Multiplexing (TDM)   9
       1.2.3   Frequency Hopping 10
       1.2.4   Direct Sequence Spread Spectrum 10
   1.3   Circuit Switching Versus Packet Switching 11
   1.4   Layered Protocols 13
   1.5   Ethernet 15
       1.5.1   10 Mbps Ethernet 15
       1.5.2   Fast Ethernet 18
       1.5.3   Gigabit Ethernet 19
       1.5.4   10 Gigabit Ethernet 21
   1.6   Wireless Networks 22
       1.6.1   802.11 WiFi 22
       1.6.2   802.15 Bluetooth 26
       1.6.3   802.16 Wireless MAN 28
   1.7   ATM 30
       1.7.1   Limitations of STM 30
       1.7.2   ATM Features 31

       1.7.3 ATM Switching 34
    1.8   SONET 36
       1.8.1 SONET Architecture 37
       1.8.2 Self-Healing Rings 39
    1.9   Wavelength Division Multiplexing (WDM)................. 40
   1.10 Grids 41
   1.11 Problems 43

2 Fundamental Stochastic Models 45
    2.1   Introduction 45
    2.2   Bernoulli and Poisson Processes 46
    2.3   Bernoulli Process Statistics 52
    2.4   Multiple Access Performance............................. 57
       2.4.1 Introduction 57
       2.4.2 Discrete Time Ethernet Model 57
       2.4.3 Ethernet Design Equation 60
       2.4.4 Aloha Multiple Access Throughput Analysis 62
       2.4.5 Aloha Multiple Access Delay Analysis................ 66
    2.5   Teletraffic Modeling for Specific Topologies 69
       2.5.1 Introduction 69
       2.5.2 Linear Network 69
       2.5.3 Tree Networks 71
       2.5.4 Two-dimensional Circular Network 75
    2.6   Switching Elements and Fabrics 78
       2.6.1 Introduction 78
       2.6.2 Switching Elements 79
       2.6.3 Networks 84
    2.7   Conclusion 92
    2.8   Problems 92

3 Queueing Models 99
    3.1   Introduction 99
    3.2   Single Queue Models...................................... 100
       3.2.1 M/M/1 Queue 100
       3.2.2 Geom/Geom/1 Queue 108
    3.3   Some Important Single Queue Models...................... 113
       3.3.1 The Finite Buffer M/M/1 Queueing System 113
       3.3.2 The M/M/m/m Loss Queueing System 114
       3.3.3 M/M/m Queueing System 117
       3.3.4 A Queueing-Based Memory Model................... 120
       3.3.5 M/G/1 Queueing System........................ 122
    3.4   Common Performance Measures 126
    3.5   Markovian Queueing Networks 127
       3.5.1 Open Networks 128
       3.5.2 Closed Networks 132

3.6   Mean Value Analysis for Closed Networks 134
      3.6.1   MVA for Cyclic Networks 135
      3.6.2   MVA for Random Routing Networks 138
3.7   Negative Customer Queueing Networks..................... 140
      3.7.1   Negative Customer Product Form Solution 141
3.8   Recursive Solutions for State Probabilities 144
3.9   Stochastic Petri Nets 148
      3.9.1   Petri Net Schematics 148
      3.9.2   Petri Net Markov Chains 149
3.10 Solution Techniques 152
      3.10.1 Analytical Solutions 152
      3.10.2 Numerical Computation 152
      3.10.3 Simulation..................................... 153
3.11 Conclusion 154
3.12 Problems 155

4   **Fundamental Deterministic Algorithms** 161
4.1   Introduction 161
4.2   Routing 161
      4.2.1   Introduction 161
      4.2.2   Dijkstra's Algorithm 163
      4.2.3   Ford Fulkerson Algorithm 166
      4.2.4   Table Driven Routing 167
      4.2.5   Source Routing 168
      4.2.6   Flooding 169
      4.2.7   Hierarchical Routing 169
      4.2.8   Self-Routing 170
      4.2.9   Multicasting 173
      4.2.10 Ad Hoc Network Routing 173
4.3   Protocol Verification 174
4.4   Error Codes 178
      4.4.1   Introduction 178
      4.4.2   Parity Codes.................................... 180
      4.4.3   Hamming Error Correction 181
      4.4.4   The CRC Code 184
4.5   Conclusion 188
4.6   Problems 189

5   **Divisible Load Modeling for Grids** 193
5.1   Introduction 193
5.2   Some Single Level Tree (Star) Networks 200
      5.2.1   Sequential Load Distribution...................... 201
      5.2.2   Simultaneous Distribution, Staggered Start 205
      5.2.3   Simultaneous Distribution, Simultaneous Start 210
      5.2.4   Nonlinear Load Processing Complexity 215

5.3   Equivalent Processors 221
    5.3.1   The Tree Network Without Front-End Processors 222
    5.3.2   The Tree Network With Front-End Processors 230
5.4   Infinite-Sized Network Performance 236
    5.4.1   Linear Daisy Chains 236
    5.4.2   Tree Networks 245
5.5   Time-Varying Environments 248
5.6   Linear Programming and Divisible Load Modeling 255
5.7   Experimental Work 257
5.8   Conclusion 259
5.9   Problems 259

A   Summation Formulas 263

References 265

Index 275

1

A Tour through Networking and Grids

Something about technology allows people and their computers to communicate with each other, which makes networking a fascinating field, both technically and intellectually.

What is a network? It is a collection of computers (nodes) and transmission channels (links) that allow people to communicate over distances, large and small. A Bluetooth personal area network may simply connect your home PC with its peripherals. An undersea fiber-optic cable may traverse an ocean. The Internet and telephone networks span the globe.

What is a grid? A grid is a special type of network integrated with (usually powerful) computers and storage devices to give a user located anywhere on the globe the ability to have a virtual worldwide computer on which they can run (often massive) jobs. The scientific community has been in the forefront of grid development efforts. They are interested in distributed processing of large scientific jobs in such diverse fields as geology, environmental science, physics, and astronomy.

Networking in particular has been a child of the late twentieth century. The Internet has been developed over the past 35 years. The 1980s and 1990s saw the birth and growth of local area networks, synchronous optical networking (SONET) fiber networks, and asynchronous transfer mode (ATM) backbones. The 1990s and the early years of the new century have seen the development and expansion of wavelength division multiplexing (WDM) fiber multiplexing and grids.

The purpose of this chapter is to give a concise overview of some major topics in networking and grids. The succeeding chapters examine key mathematical approaches and techniques for networking and grids.

Chapter 2 covers fundamental stochastic (i.e., random) models for networking. Chapter 3 provides an introduction to queueing theory, a widely used tool for modeling and predicting the performance of networked systems. In chapter 4, some fundamental deterministic algorithms for networking are studied. These algorithms include shortest path routing, protocol verification, and error checking codes. Finally, chapter 5 provides an extensive tutorial on divisible load scheduling theory, a relatively new performance evaluation methodology that has applications to grid computing.

We now start our one-chapter tour through the applied aspects of networks and grids by discussing the means of achieving connectivity.

1.1 Achieving Connectivity

A variety of transmission methods, both wired and wireless, are available today to provide connectivity between computers, networks, and people. Wired transmission media include coaxial cable, twisted pair wiring, and fiber optics. Wireless technology includes microwave line of sight, satellites, cellular systems, ad hoc networks, and wireless sensor networks. We now review these media and technologies.

1.1.1 Coaxial Cable

You may have this thick cable in your house to connect your cable TV setup box to the outside wiring plant. This type of cable has been around for many years and is a mature technology. Although still popular for cable TV systems today, it was also a popular choice for wiring local area networks in the 1980s. It was used in the wiring of the original 10 Mbps Ethernet.

A coaxial cable has four parts: a copper inner core, surrounded by insulating material, a metallic outer conductor, and a plastic outer cover. Essentially, in a coaxial cable, there are two wires (copper inner core and outer conductor) with one geometrically inside the other. This configuration reduces interference to/from the coaxial cable with respect to other nearby wires.

The bandwidth of a coaxial cable is on the order of 1 GHz. How many bits per second can it carry? Modulation is used to match a digital stream to the spectrum-carrying ability of the cable. Depending on the efficiency of the modulation scheme used, 1 bps requires anywhere from 1/14 to 4 Hz. For short distances, a coaxial cable may use 8 bits/Hz or carry 8 Gbps.

Different types of coaxial cable also exist. One type with a 50 ohm termination is used for digital transmissions. Another type with a 75 ohm termination is used for analog transmissions or cable TV systems.

A word is in order on cable TV systems. Such networks are locally wired as tree networks with the root node called the head end. At the head end, programming is brought in by fiber or satellite. The head end may also have a TV studio for public access programming. From the head end, cables (and possibly fiber) radiate out to homes. Amplifiers may be placed in this network when distances are large.

For many years, cable TV companies were interested in providing two-way service. Although early limited trials were generally not successful (except for Video on Demand), recently cable TV seems to have winners in broadband access to the Internet and in carrying telephone traffic.

1.1.2 Twisted Pair Wiring

Coaxial cable is generally no longer used for wiring local area networks. One type of replacement wiring has been twisted pair. Twisted pair wiring typically had been previously used to wire phones to the telephone network. A twisted pair consists of two wires twisted together over their length. The twisted geometry reduces electromagnetic leakage (i.e., cross-talk) with nearby wires. Twisted pairs can run several kilometers without the need for amplifiers. The quality of a twisted pair (carrying capacity) depends on the number of twists per inch.

About 1990, it became possible to send 10 Mbps (for Ethernet) over unshielded twisted pair (UTP). Higher speeds are also possible if the cable and connector parameters are carefully implemented.

One type of unshielded twisted pair is category 3 UTP. It consists of four pairs of twisted pair surrounded by a sheath. It has a bandwidth of 16 MHz. Many offices used to be wired with category 3 wiring.

Category 5 UTP has more twists per inch. Thus, it has a higher bandwidth (100 MHz). Up and coming standards include category 6 (250 MHz) and category 7 (600 MHz). Shielded twisted pair is also possible but has not been used much beyond IBM equipment.

The fact that twisted pair is lighter and thinner than coaxial cable has speeded its widespread acceptance.

1.1.3 Fiber Optics

Fiber-optic cable consists of a silicon glass core that conducts light, rather than electricity, as in coaxial cables and twisted pair wiring. The core is surrounded by cladding and then a plastic jacket.

Fiber-optic cables have the highest data-carrying capacity of any wired medium. A typical fiber has a capacity of 50 Tbps (terabits per second or 50×10^{12} bits per second). In fact, this data rate for years has been much higher than the speed at which standard electronics could load the fiber. This mismatch between fiber and nodal electronics speed has been called the "electronic bottleneck." Decades ago the situation was reversed, links were slow, and nodes were relatively fast. This paradigm shift has led to a redesign of protocols.

Two major types of fiber exist: multi-mode and single mode. Pulse shapes are more accurately preserved in single-mode fiber, lending to a higher potential data rate. However, the cost of multi-mode and single-mode fiber is comparable. The real difference in pricing is in the opto-electronics needed at each end of the fiber. One reason multi-mode fibers have a lower performance is dispersion. Under dispersion, square digital pulses tend to spread out in time, thus lowering the potential data rate. Special pulse shapes (such as hyperbolic cosines) called solitons, for which dispersion is minimized, have been the subject of research.

Mechanical fiber connectors to connect two fibers can lose 10% of the light that the fiber carries. Fusing two ends of the fiber results in a smaller attenuation.

Fiber-optic cables today span continents and are laid across the bottom of oceans between continents. They are also used by organizations to internally carry telephone, data, and video traffic.

1.1.4 Microwave Line of Sight

Microwave radio energy travels largely in straight lines. Thus, some network operators construct networks of tall towers kilometers apart and place microwave antennas at different heights on each tower. Although the advantage is that no need exists to dig trenches for cables, the expense of tower construction and maintenance must be taken into account.

1.1.5 Satellites

Arthur C. Clarke, the science fiction writer, first proposed using satellites as communication relays in the late 1940s. Satellites are now extensively used for communication purposes. They fill certain technological niches very well: providing connectivity to mobile users, for large area broadcasts, and for communications for areas with poor infrastructure. The two main communication satellite architectures are geostationary satellites and low Earth orbit satellites (LEOS). Both are now discussed.

Geostationary Satellites

You may recall from a physics course that a satellite in a low orbit (hundreds of kilometers) around the equator seems to move against the sky. As its orbital altitude increases, its apparent movement slows. At a certain altitude of approximately 36,000 km, it appears to stay in one spot in the sky, over the equator, 24 hours a day. In reality, the satellite is moving around the Earth but at the same angular speed that the Earth is rotating, giving the illusion that it is hovering in the sky.

This comparison is very useful. For instance, a satellite TV service can install home antennas that simply point to the spot in the sky where the satellite is located. Alternatively, a geostationary satellite can broadcast a signal to a large area (its "footprint") 24 hours a day.

By international agreement, geostationary satellites are placed 2 degrees apart around the equator. Some locations are more economically preferable than others, depending on which regions of the Earth are under the location.

A typical geostationary satellite will have several dozen transponders (relay amplifiers), each with a bandwidth of 80 MHz (Tanenbaum 03). Such a satellite may weigh several thousand kilograms and consume several kilowatts using solar panels.

The number of microwave frequency bands used have increased over the years as the lower bands have become crowded and technology has improved. Frequency bands include L (1.5/1.6 GHz), S (1.9/2.2 GHz), C (4/6 GHz), Ku (11/14 GHz), and Ka (20/30 GHz) bands. Here the first number is the downlink band and the second number is the uplink band. The actual bandwidth of a signal may vary from about 15 MHz in the L band to several GHz in the Ka band (Tanenbaum 03).

It should be noted that extensive studies of satellite signal propagation under different weather and atmospheric conditions have been conducted. Excess power for overcoming rain attenuation is often budgeted above 11 GHz.

Low Earth Orbit Satellites

A more recent architecture is that of LEOS. The most famous LEOS system was Iridium from Motorola. It received its name because the original proposed 77-satellite network has the same number of satellites as the atomic number of the element Iridium. In fact, the actual system orbited had 66 satellites, but the system name Iridium was kept.

The purpose of Iridium was to provide a global cell phone service. One would be able to use an Iridium phone anywhere in the world (even on the ocean or in the Artic). Unfortunately, after spending $5 billion to deploy the system, talking on Iridium cost a dollar or more a minute while local terrestrial cell phone service was under 25 cents a minute. Although an effort was made to appeal to business travelers, the system was not profitable, eventually closed, and sold. Another problem was that every several years or so the satellites would deorbit and have to be replaced.

Technologically, though, the Iridium system is interesting. Eleven satellites exist in each of six polar orbits (passing over the North Pole, south to the South Pole, and back up to the North Pole; see Figure 1.1).

At any given time, several satellites are moving across the sky over any location on Earth. Using several dozen spot beams, the system can support almost a quarter of a million conversations. Calls can be relayed from satellite to satellite.

It should be noted that, when Iridium was hot, several competitors were proposed but not built. One competitor used a "bent pipe" architecture, where a call to a satellite would be beamed down from the same satellite to a ground station and then sent over the terrestrial phone network rather than be relayed from satellite to satellite. This architecture was built in an effort to lower costs and simplify the design.

1.1.6 Cellular Systems

Starting around the early 1980s, cellular telephone systems, which provide connectivity between mobile phones and the public switched telephone network, were deployed. In such systems, signals go from/to a cell phone to/from

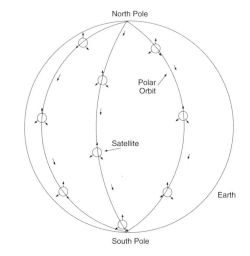

Fig. 1.1. LEOS in polar orbits

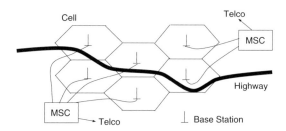

Fig. 1.2. Part of a cellular network

a local "base station" antenna, which is hardwired into the public switched telephone network. Figure 1.2 illustrates such a system. A geographic region such as a city or suburb is divided into geographic sub-regions called "cells."

Base stations are shown at the center of cells. Nearby base stations are wired into a switching computer (the mobile switching center or MSC) that provides a path to the telephone network.

A cell phone making a call connects to the nearest base station (i.e., the base station with the strongest signal). Base stations and cell phones measure and communicate received power levels. If one is driving and one approaches a new base station, the cell phone signal will at some point become stronger than that of the original base station one is connected to and the system will then perform a "handoff." In a handoff, connectivity is changed from one base station to an adjacent one. Handoffs are transparent; the talking user is not aware when one occurs.

Calls to a cell phone involve a paging-like mechanism that activates (rings) the called user's phone.

The first cellular system was deployed in 1979 in Japan by NTT. The first U.S. cellular system was AMPS (Advanced Mobile Phone System) from AT&T. It was first deployed in 1983. These phones were first-generation analog systems. Second-generation systems were digital. The most popular is the European originated GSM (Global System for Mobile), which has been installed all over the world. Third- and fourth-generation cellular systems provide increased data rates for such applications as Internet browsing and picture transmission.

1.1.7 Ad Hoc Networks

Ad hoc networks are radio networks where (often mobile) nodes can come together, transparently form a network without any user interaction, and maintain the network as long as the nodes are in range of each other and energy supplies last (Rabaey 00, Mauve 01). In an ad hoc network, messages hop from node to node to reach an ultimate destination. For this reason ad hoc networks used to be called multi-hop radio networks. In fact, because of the nonlinear dependence of energy on transmission distance, the use of several small hops uses much less energy than a single large hop, often by orders of magnitude.

Ad hoc network characteristics include multi-hop transmission, possibly mobility and possibly limited energy to power the network nodes. Applications include mobile networks, emergency networks, wireless sensor networks, and ad hoc gatherings of people, as at a convention center.

Routing is an important issue for ad hoc networks. Two major categories of routing algorithms are topology-based routing and position-based routing. Topology-based routing uses information on current links to perform the routing. Position-based routing makes use of a knowledge of the geographic location of each node to route. The position information may be acquired from a service such as the global positioning system (GPS).

Topology-based algorithms may be further divided into proactive and reactive algorithms. Proactive algorithms use information on current paths as inputs to classic routing algorithms. However, to keep this information current, a large amount of control message traffic is needed, even if a path is unused. This overhead problem is exacerbated if there are many topology changes (say due to movement of the nodes).

On the other hand, reactive algorithms such as DSR, TORA, and AODV maintain routes only for paths currently in use to keep the amount of information and control overhead more manageable. Still, more control traffic is generated if there are many topology changes.

Position-based routing does not require maintenance of routes, routing tables, or generation of large amounts of control traffic other than information regarding positions. "Geocasting" to a specific area can be simply implemented. Several heuristics can be used in implementing position-based routing.

1.1.8 Wireless Sensor Networks

The integration of wireless, computer, and sensor technology has the potential to make possible networks of miniature elements that can acquire sensor data and transmit the data to a human observer. Wireless sensor networks are receiving an increasing amount of attention from researchers in universities, government, and industry because of their promise to become a revolutionary technology and the technical challenges that must be overcome to make this a reality. It is assumed that such wireless sensor networks will use ad hoc radio networks to forward data in a multi-hop mode of operation.

Typical parameters for a wireless sensor unit (including computation and networking circuitry) include a size from 1 millimeter to 1 centimeter, a weight less than 100 grams, cost less than $1 and power consumption less than 100 microwatts (Shah 02). By way of contrast, a wireless personal area network Bluetooth transceiver consumes more than a 1000 microwatts and costs more than $10. A cubic millimeter wireless sensor can store, with battery technology, 1 Joule allowing a 10 microwatt energy consumption for 1 day (Kahn 00). Thus, energy scavenging from light or vibration has been proposed. Note also that data rates are often relatively low for sensor data (100s bps to 100 Kbps).

Naturally, with these parameters, minimizing energy usage in wireless sensor networks becomes important. Although in some applications wireless sensor networks may be needed for a day or less, there are many applications where a continuous source of power is necessary. Moreover, communication is much more energy expensive than computation. Sending one bit for a distance of 100 meters can take as much energy as processing 3000 instructions on a micro-processor.

Although military applications of wireless sensor networks are fairly obvious, there are many potential scientific and civilian applications of wireless sensor networks. Scientific applications include geophysical, environmental, and planetary exploration. One can imagine wireless sensor networks being used to investigate volcanos, measure weather, monitor beach pollution, or record planetary surface conditions.

Biomedical applications include applications such as glucose level monitoring and retinal prosthesis (Schwiebert 01). Such applications are particularly demanding in terms of manufacturing sensors that can survive in and not affect the human body.

Sensors can be placed in machines (where vibration can sometimes supply energy) such as rotating machines, semiconductor processing chambers, robots, and engines. Wireless sensors in engines could be used for pollution control telemetry.

Finally, among many potential applications, wireless sensors could be placed in homes and buildings for climate control. Note that wiring a single sensor in a building can cost over $200 plus the sensor cost (Rabaey 00). Ultimately, wireless sensors could be embedded in building materials.

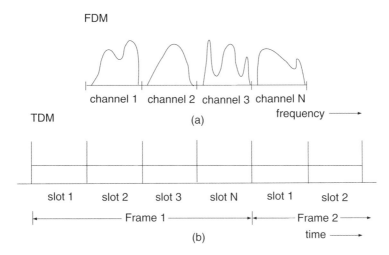

Fig. 1.3. (a) Frequency division multiplexing and (b) time division multiplexing

1.2 Multiplexing

Multiplexing involves sending multiple signals over a single medium. Thomas Edison invented a four-to-one telegraph multiplexer that allowed four telegraph signals to be sent over one wire. The major forms of multiplexing for networking today are frequency division multiplexing (FDM), time division multiplexing (TDM), and spread spectrum. Each is now reviewed.

1.2.1 Frequency Division Multiplexing (FDM)

Here a portion of spectrum (i.e., band of frequencies) is reserved for each channel [Figure 1.3(a)]. All channels are transmitted simultaneously, but a tunable filter at the receiver only allows one channel at a time to be received. This is how AM, FM, and television signals are transmitted. Moreover, it is how distinct optical signals are transmitted over a fiber using WDM technology.

1.2.2 Time Division Multiplexing (TDM)

Time division multiplexing is a digital technology that, on a serial link, breaks time into equi-duration slots [Figure 1.3(b)]. A slot may hold a voice sample in a telephone system or a packet in a packet switching system. A frame consists of N slots. Frames, and thus slots, repeat. A telephone channel might use slot 14 of 24 slots in a frame during the length of a call, for instance.

Time division multiplexing is used in the second-generation cellular system, GSM. It is also used in digital telephone switches. Such switches in fact use electronic devices called time slot interchangers that transfer voice samples from one slot to another to accomplish switching.

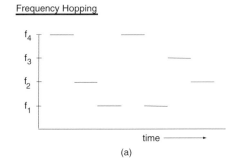

(a)

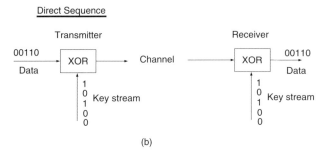

(b)

Fig. 1.4. (a) Frequency hopping spread spectrum and (b) direct sequence spread spectrum

1.2.3 Frequency Hopping

Frequency hopping is one form of spread spectrum technology and is typically used on radio channels. The carrier (center) frequency of a transmission is pseudo-randomly hopped among a number of frequencies [Figure 1.4(a)]. The hopping is done in a deterministic, but random-looking, pattern that is known to both transmitter and receiver. If the hopping pattern is known only to the transmitter and receiver, one has good security. Frequency hopping also provides good interference rejection. Multiple transmissions can be multiplexed in the same local region if each uses a sufficiently different hopping pattern. Frequency hopping dates back to the era of World War II.

1.2.4 Direct Sequence Spread Spectrum

This alternative spread spectrum technology uses exclusive or (xor) gates as scramblers and de-scramblers [Figure 1.4(b)]. At the transmitter, data are fed into one input of an xor gate and a pseudo-random key stream into the other input.

From the xor truth table, one can see that, if the key bit is a zero, the output bit equals the data bit. If the key bit is a one, the output bit is the

Table 1.1. XOR Truth Table

Key	Data	Output
0	0	0
0	1	1
1	0	1
1	1	0

complement of the data bit (0 becomes 1, 1 becomes 0). This scrambling action is quite strong under the proper conditions. Unscrambling can be performed by an xor gate at the receiver. The transmitter and receiver must use the same (synchronized) key stream for this to work. Again, multiple transmissions can be multiplexed in a local region if the key streams used for each transmission are sufficiently different.

1.3 Circuit Switching Versus Packet Switching

Two major architectures for networking and telecommunications are circuit switching and packet switching. Circuit switching is the older technology, going back to the years following the invention of the telephone in the late 1800s. As illustrated in Figure 1.5(a), for a telephone network, when a call has to be made from node A to node Z, a physical path with appropriate resources called a "circuit" is established. Resources include link bandwidth and switching resources. Establishing a circuit requires some setup time before actual communication commences. Even if one momentarily stops talking, the circuit is still in operation. When the call is finished, link and switching resources are released for use by other calls. If insufficient resources are available to set up a call, the call is said to be blocked.

Packet switching was created during the 1960s. A packet is a bundle of bits consisting of header bits and payload bits. The header contains the source and destination address, priority levels, error check bits, and any other information that is needed. The payload is the actual information (data) to be transported. However, many packet switching systems have a maximum packet size. Thus, larger transmissions are split into many packets and the transmission is reconstituted at the receiver.

The diagram of Figure 1.5(b) shows packets, possibly from the same transmission, taking multiple routes from node A to node Z, which is called datagram or connectionless-oriented service. Packets may indeed take different routes in this type of service as nodal routing tables are updated periodically in the middle of a transmission.

A hybrid type of service is the use of "virtual circuits" or connection-oriented service. Here packets belonging to the same transmission are forced to take the same serial path through the network. A virtual circuit has an identification number that is used at nodes to continue the circuit along its

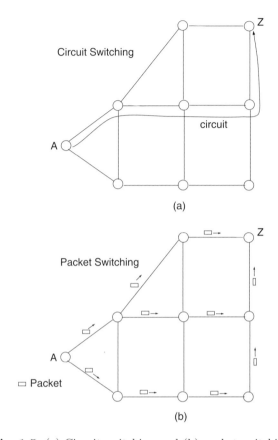

Fig. 1.5. (a) Circuit switching and (b) packet switching

preset path. As in circuit switching, a virtual circuit needs to be set up prior to its use for communication. That is, entries need to be made in routing tables (see chapter 4) implementing the virtual circuit.

An advantage of virtual circuit usage is that packets arrive at the destination in the same order that they were sent, which avoids the need for buffers for reassembling transmissions (reassembly buffers) that are needed when packets arriving at the destination are not in order, as in datagram service. As we shall see, ATM, the high-speed packet switching technology used in Internet backbones, uses virtual circuits.

Packet switching is advantageous when traffic is bursty (occurs at irregular intervals) and individual transmissions are short. It is a very efficient way of sharing network transmissions when there are many such transmissions. Circuit switching is not well suited for bursty and short transmissions. It is more efficacious when transmissions are relatively long (to minimize setup time overhead) and provide a constant traffic rate (to well utilize the dedicated circuit resource).

1.4 Layered Protocols

Protocols are the rules of operation of a network. A common way to engineer a complex system is to break it into more manageable and coherent components. Network protocols are often divided into layers in the layered protocol approach. Figure 1.6 illustrates the general OSI (open systems interconnection) protocol stack. Proprietary protocols may have different names for the layers and/or a different layer organization, but pretty much all networking protocols have the same functionality.

Transmissions in a layered architecture (see Figure 1.6) move from the source's top layer (application), down the stack to the physical layer, through a physical channel in a network, to the destination's physical layer, up the destination stack to the destination application layer. Note that any communication between peer layers must move down one stack, across and up the receiver's stack. It should also be noted that, if a transmission passes through an intermediate node, only some lower layers (e.g., network, data link, and physical) may be used at the intermediate nodes.

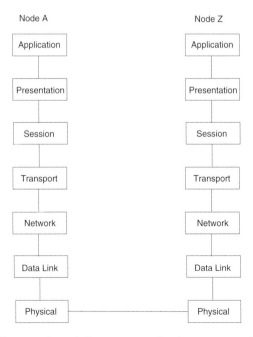

Fig. 1.6. OSI protocol stack for a communicating source and destination

It is interesting that a packet moving down the source's stack may have its header grow as each layer may append information to the header. At the destination, each layer may remove information from the packet header, which causes it to decrease in size as it moves up the stack.

In a particular implementation, some layers may be larger and more complex, whereas others are relatively simple.

In the following discussion, we briefly discuss each layer.

Application Layer

Traditional applications for networking include e-mail, remote login, file transfer and more recently, the World Wide Web. But an application may also be more specialized, such as distributed software to run a network of catalog company order depots.

Presentation Layer

This layer controls how information is formatted, such as on a screen (number of lines, number of characters across).

Session Layer

This layer is important for managing a session, as in remote logins. In other cases, this is not a concern.

Transport Layer

This layer can be thought of as an interface between the upper and the lower layers. More importantly, it is designed to give the impression to the layers above that they are dealing with a reliable network, even though the layers below the transport layer may not be perfectly reliable. For this reason, some think of the transport layer as the most important layer.

Network Layer

The network layer manages multiple links. Its most important function is to do routing. Routing involves selecting the best path for a circuit or packet stream. Routing algorithms are discussed in chapter 4.

Data Link Layer

Whereas the network layer manages multiple link functions, a data link protocol manages a single link. One of its potential functions is encryption, which can either be done on a link-by-link basis (i.e., at the data link layer) or on an end-to-end basis (i.e., at the transport layer) or both. End-to-end encryption is a more conservative choice as one is never sure what type of subnetwork a transmission may pass through and what its level of encryption is, if any.

Physical Layer

The physical layer is concerned with the raw transmission of bits. Thus, it includes engineering physical transmission media, modulation and demodulation, and radio technology. Many communication engineers work on physical layer aspects of networks. Again, the physical layer of a protocol stack is the only layer that provides actual direct connectivity to peer layers.

1.5 Ethernet

Local area networks (LANs) are networks that cover a small area as in a department in a company or university. In the early 1980s, the three major local area networks were Ethernet (IEEE Standard 802.3), Token Ring (802.5 and used extensively by IBM), and Token Bus (802.4, intended for manufacturing plants). However, over the years, Ethernet has become the most popular local area network standard. While maintaining a low cost, it has gone through four versions, each one ten times faster than the previous version (10 Mbps, 100 Mbps, 1 Gbps, 10 Gbps).

Ethernet was invented at the Xerox Palo Alto Research Center (PARC) by Metcalfe and Boggs, circa 1976. It is similar in spirit to the earlier Aloha radio protocol (see chapter 2), although the scale is smaller. IEEE's 802.3 committee produced the first Ethernet standard. Xerox never produced Ethernet commercially, but other companies did.

In going from one Ethernet version to the next, the IEEE 802.3 committee sought to make each version similar to the previous ones and to use existing technology. In the following discussion we now discuss the various versions of Ethernet.

1.5.1 10 Mbps Ethernet

Back in the 1980s, Ethernet was originally wired using coaxial cable. As in Figure 1.7(a), a coaxial cable was snaked through the floor or ceiling and computers were attached to it along its length. The coaxial cable acted as a private radio channel that each computer would monitor. If a station had a packet to send, it would send it immediately if the channel was idle. If the station sensed the channel to be busy, it would wait until the channel was free. In all of this, only one transmission can be on the channel at one time.

A problem occurs if two or more stations sense the channel to be idle at about the same time and attempt to transmit simultaneously. The packets overlap in the cable and are garbled, which is a collision. The stations involved, using analog electronics, can detect the collision, stop transmitting, and reschedule their transmissions.

Thus, the price one pays for this completely decentralized access protocol is the presence of utilization lowering collisions. The protocol used goes by the

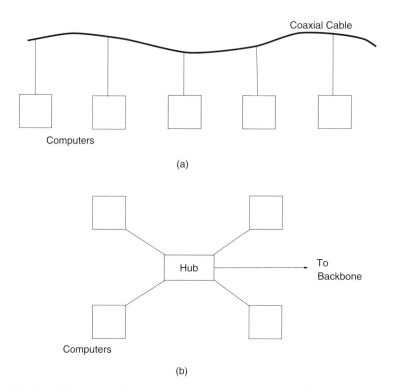

Fig. 1.7. Ethernet wiring using (a) coaxial cable and (b) hub topology

name 1-persistent CSMA/CD (Carrier Sense Multiple Access with Collision Detection). The name is pretty much self-explanatory except that 1-persistent refers to the fact that a station with a packet to send attempts this on an idle channel with a probability of 1.0. In a CSMA/CD protocol, if the bit rate is 10 Mbps, the actual useful information transport can be significantly less because of collisions (or occasional idleness).

In the case of a collision, the rescheduling algorithm used is called Binary Exponential Backoff. Under this protocol, two or more stations experiencing a collision randomly reschedule over a time window with a default of 51 microseconds for a 500 meter network. If a station becomes involved in a second collision, it doubles its window size and attempts again to randomly reschedule its transmission. Windows may be doubled in size up to ten times. Once a packet is successfully transmitted, the window size drops back to the default (smallest) value for that packet's station. Thus, this protocol at a station has no long-term memory regarding past transmissions.

Table 1.2 above shows the fields in the 10 Mbps Ethernet frame. A frame is the name for a packet at the data link layer. The preamble is for communication receiver synchronization purposes. Addresses are either local (2 bytes)

Table 1.2. Ethernet Frame Format

Field	Length
Preamble	7 bytes
Frame Delimiter	1 byte
Destination Address	2 or 6 bytes
Source Address	2 or 6 bytes
Data Length	2 bytes
Data	up to 1500 bytes
Pad	variable
CRC Checksum	4 bytes

or global (6 bytes). Note that Ethernet addresses are different from IP addresses. Different amounts of data can be accommodated up to 1500 bytes. Transmissions longer than 1500 bytes of data must be segmented into multiple packets. The pad field is used to guarantee that the frame is at least 64 bytes in length (minimum frame size) if the frame would be less than 64 bytes in length. Finally the checksum is based on CRC error detecting coding (see chapter 4).

A problem with digital receivers is that they require many 0 to 1 and 1 to 0 transitions to properly lock onto a signal. But long runs of 1s or 0s are not uncommon in data. To provide many transitions between logic levels, even if the data have a long run of one logic level, Ethernet uses Manchester encoding.

Referring to Figure 1.8, under Manchester encoding, if a logic 0 needs to be sent, a transition is made for 0 to 1 (low to high voltage) and if a logic 1 needs to be sent, the opposite transition is made for 1 to 0 (high to low voltage). The voltage level makes a return to its original level at the end of a bit as necessary. Note that the "signaling rate" is variable. That is, the number of transitions per second is twice the data rate for long runs of a logic level and is equal to the data rate if the logic level alternates. For this reason, Manchester

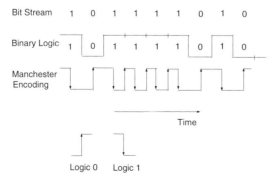

Fig. 1.8. Manchester encoding

Table 1.3. Original Ethernet Wiring

Cable	Type	Maximum Size
10Base5	Thick Coax	500 m
10Base2	Thin Coax	200 m
10Base-T	Twisted Pair	100 m
10Base-F	Fiber Optics	2 km

encoding is said to have an efficiency of 50%. More modern signaling codes, such as 4B5B, achieve 80% efficiency (see Fast Ethernet below).

During the 1980s, Ethernets were wired with linear coaxial cables. Today hubs are commonly used [Figure 1.7(b)]. These are boxes (some smaller than a cigar box) that computers tie into, in a star type wiring pattern, with the hub at the center of the star.

A hub may internally have multiple cards, each of which have multiple external Ethernet connections. A high-speed (in the gigabits) proprietary bus interconnects the cards. Cards may mimic a CSMA/CD Ethernet with collisions (shared hub) or use buffers at each input (switched hub). In a switched hub, multiple packets may be received simultaneously without collisions, raising throughput.

The next table (Table 1.3) illustrates Ethernet wiring. In "10 Base5," the 10 stands for 10 Mbps and the 5 for the 500 meter maximum size. Used in the early 1980s, 10 Base5 used vampire taps that would puncture the cable. Also, at the time, 10 Base2 used T junctions and BNC connectors as wiring hardware. Today, 10 Base-T is the most common wiring solution for 10 Mbps Ethernet. Fiber optics, 10 Base-F, is only intended for runs between buildings, but a higher data rate protocol would probably be used today for this purpose.

1.5.2 Fast Ethernet

As the original 10 Mbps Ethernet became popular and the years passed, traffic on Ethernet networks continued to grow. To maintain performance, network administrators were forced to segment Ethernet networks into smaller networks (each handling a smaller number of computers) connected by a spaghetti-like arrangement of repeaters, bridges, and routers. In 1992, IEEE assigned the 802.3 committee the task of developing a faster local area network protocol.

The committee agreed on a 100 Mbps protocol that would incorporate as much of the existing Ethernet protocol/technology as possible to gain acceptance and so that they could move quickly. The resulting protocol, IEEE 802.3u, was called Fast Ethernet.

Fast Ethernet is only implemented with hubs, in a star topology [Figure 1.7(b)]. There are three major wiring options (Table 1.4).

Table 1.4. Fast Ethernet Wiring

Cable	Type	Maximum Size
100Base-T4	Twisted Pair	100 m
100Base-TX	Twisted Pair	100 m
100Base-FX	Fiber Optics	2 km

The original Ethernet has a data rate of 10 Mbps and a maximum signaling rate of 20 MHz (recall that the Manchester encoding used was 50% efficient). Fast Ethernet 100 Base-T4 with its data rate of 100 Mbps has a signaling speed of 25 MHz, not 200 MHz. How is this accomplished?

Fast Ethernet 100 Base-T4 actually uses four twisted pairs per cable. Three twisted pairs carry signals from its hub to a PC. Each of the three twisted pairs uses ternary (not binary) signaling using three logic levels. Thus, one of $3 \times 3 \times 3 = 27$ symbols can be sent at once. Only 16 symbols are used, though, which is equivalent to sending 4 bits at once. With 25 MHz clocking, 25 MHz×4 bits yields a data rate of 100 Mbps. The channel from the PC to hub operates at 33 MHz. For most PC applications, an asymmetrical connection with more capacity from hub to PC for downloads is acceptable. Category 3 or 5 unshielded twisted pair wiring is used for 100 Base-T4.

An alternative to 100 Base-T4 is 100 Base-TX. This uses two twisted pairs, with 100 Mbps in each direction. However, 100 Base-T4 has a signaling rate of only 125 MHz. It accomplishes this using 4B5B (Four Bit Five Bit) encoding rather than Manchester encoding. Under 4B5B, every four bits is mapped into five bits in such a way that there are many transitions for digital receivers to lock onto, irrespective of the actual data stream. Since four bits are mapped into five bits, 4B5B is 80% efficient. Thus, 125 MHz times 0.8 yields 100 Mbps.

Finally, 100 Base-FX uses two strands of the lower performing multi-mode fiber. It has 100 Mbps in both directions and is for runs (say between buildings) of up to 2 km.

It should be noted that Fast Ethernet uses the signaling method for twisted pair (for 100 Base-TX) and fiber (100 Base-FX) borrowed from fiber distributed data interface (FDDI). The FDDI protocol was a 100 Mbps token ring protocol used as a backbone in the 1980s.

To maintain channel efficiency (utilization) at 100 Mbps, versus the original 10 Mbps, the maximum network size of Fast Ethernet is about ten times smaller than that of the original Ethernet. See chapter 2 for an analysis of the relevant Ethernet design equation.

1.5.3 Gigabit Ethernet

The ever growing amount of network traffic brought on by the growth of applications and more powerful computers motivated a revised, faster version of Ethernet. Approved in 1998, the next version of Ethernet operates at

1000 Mbps or 1 Gbps and is known as Gigabit Ethernet, or 802.3z. As much as possible, the Ethernet committee sought to utilize existing Ethernet features.

Gigabit Ethernet wiring is either between two computers directly or, as is more common, in a star topology with a hub or switch in the center of the star. In this connection, it is appropriate to say something about the distinction between a hub and a switch. A shared medium hub uses the established CSMA/CD protocol so collisions can occur. At most, one attached station can successfully transmit through the hub at a time, as one would expect with CSMA/CD. The half duplex Gigabit Ethernet mode uses shared medium hubs.

A switch, on the other hand, does not use CSMA/CD. Rather, the use of buffers means multiple attached stations may send and receive distinct communications to/from the switch at the same time. The use of multiple simultaneous transmissions means that switch throughput is substantially greater than that of a single input line. Level 2 switches are usually implemented in software, level 3 switches implement routing functions in hardware (Stallings 02). Full duplex Gigabit Ethernet most often uses switches.

In terms of wiring, Gigabit Ethernet has two fiber-optic options (1000 Base-SX and 1000 Base-LX), a copper option (1000 Base-CX) and a twisted pair option (1000-Base-T).

The Gigabit Ethernet fiber option deserves some comment. It makes use of 8B10B encoding, which is similar in its operation to Fast Ethernet's 4B5B. Under 8B10B, 8 bits (1 byte) are mapped into 10 bits. The extra redundancy this involves allows each 10 bits not to have an excessive number of bits of the same type in a row or too many bits of one type in each of 10 bits. Thus, there are sufficient transitions from 1 to 0 and 0 to 1 or the data stream even if the data have a long run of 1s and 0s.

Gigabit Ethernet using twisted pair uses five logic levels on each wire. Four logic levels convey data, and the fifth is for control signaling. With four data logic levels, two bits are communicated at once or 8 bits over all four wires at a time. Thus, the signaling rate is 1 Gbps/8 or 125 MHz.

In terms of utilization under CSMA/CD operation, if the maximum segment size had been reduced by a factor of 10 as was done in going from the original Ethernet to Fast Ethernet, only very small gigabit networks could have been supported. To compensate for the ten times increase in data rate relative to Fast Ethernet, the minimum frame size for Gigabit Ethernet was increased (by a factor of eight) to 512 bytes from Fast Ethernet's 512 bits (see chapter 2).

Another technique that helps Gigabit Ethernet's efficiency is frame bursting. Under frame bursting, a series of frames is sent in a single burst.

Gigabit Ethernet's range is at least 500 meters for most fiber options and about 200 meters for twisted pair (Tanenbaum 03, Stallings 02).

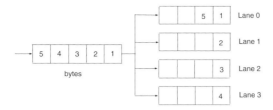

Fig. 1.9. Four parallel lanes for 10 Gigabit Ethernet

1.5.4 10 Gigabit Ethernet

Considering the improvement in Ethernet data rate over the years, it is not too surprising that a 10 Gbps Ethernet was developed (Siwamogsatham 99, Vaughan-Nichols 02). Continuing the increases in data rate by a factor of ten that have characterized the Ethernet standards, 10 Gbps (or 10,000 Mbps) Ethernet is ten times faster than Gigabit Ethernet. Applications are expected to include backbones, campus size networks, and metropolitan and wide area networks. This latter application is aided by the fact that the 10 Gbps data rate is comparable with a basic SONET fiber-optic transmission standard rate. In fact, 10 Gbps Ethernet will be a competitor to ATM high-speed packet switching technology. See the sections below for more information on SONET and ATM.

There are eight implementations of 10 Gbps Ethernet. It can use four transceiver types (one four-wavelength parallel system and three serial systems with a number of multi-mode and single mode fiber options). Like earlier versions of Ethernet, it uses CRC error coding (see chapter 4). It operates in full-duplex non-CSMA/CD mode. It can go more than 40 km via single-mode fiber.

To lower the speed at which the MAC (Media Access Control) layer processes the data stream, the MAC operates in parallel on four 2.5 Gbps streams (lanes). As illustrated in Figure 1.9, bytes in an arriving 10 Gbps serial transmission are placed in parallel in the four lanes.

A 12 byte Inter Packet Gap (IPG) is the minimum gap between packets. Normally, it would not be easy to predict the ending byte lane of the previous packet, so it would be difficult to determine the starting lane of the next transmission. The solution is to have a starting byte in a packet always occupy lane 0. The IPG is found using a pad (add in extra 1 to 3 bytes), a shrink (subtract 1 to 3 bytes), or through combination averaging (average of 12 bytes achieved through a combination of pads and shrinks). Note that padding introduces extra overhead in some implementations.

In terms of the protocol stack, this can be visualized as in Figure 1.10.

The PCS, PMA, and PMD sublayers use parallel lanes for processing. In terms of the sub-layers, they are as follows:

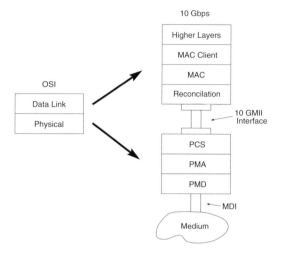

Fig. 1.10. Protocol stack for 10 Gbps Ethernet

Reconciliation: Command translator that maps terminology and commands in MAC into electrical format appropriate for physical layer.

PCS: Physical coding sublayer.

PMA: Physical medium attachment (at transmitter serialize code groups into bit stream, at receiver synchronization for data decoding).

PMD: Physical medium-dependent (includes amplification, modulation, wave shaping).

MDI: Medium-dependent interface (i.e., connector).

For more detailed discussion of the various versions of Ethernet, see Tanenbaum 03 or Stallings 02.

1.6 Wireless Networks

Wireless technology has unique capabilities to service mobile nodes and establish network infrastructure without wiring. Wireless technology has received an increasing amount of R&D attention in recent years. In this section, the popular 802.11 WiFi and 802.15 Bluetooth standards as well as the 802.16 standard are examined.

1.6.1 802.11 WiFi

The IEEE 802.11 standards (Goldberg 97, Kapp 02, LaMaire 96) have a history that goes back several years. The original standard was 802.11 (circa 1997). However, it was not that big of a marketing success because of a relatively low data rate and relatively high cost. Future standardized products

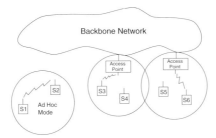

Fig. 1.11. Modes of operation for 802.11 protocol

(such as 802.11b, 802.11a, and 802.11g) were more capable and much more successful. We will start by discussing the original 802.11 standard. All 802.11 versions are meant to be wireless local area networks with ranges of several hundred feet.

The Original 802.11 Standard

The original 802.11 standard can operate in two modes (see Figure 1.11). In one mode, 802.11-capable stations connect to access points that are wired into a backbone. The other mode, ad hoc mode, allows one 802.11-capable station to connect directly to another without using an access point.

The 802.11 standard uses part of the ISM (Industrial, Scientific and Medical) band. The ISM band allows unlicensed use, unlike most other spectra. It has been popular for garage door openers, cordless telephones, and other consumer electronic devices. The ISM band includes 902–928 MHz, 2.400–2.4835 GHz, and 5.725–5.850 GHz. The original 802.11 standard used the 2.400–2.4835 GHz band.

In fact, infrared wireless local area networks have also been built but are not used today on a large scale. Using pulse position modulation (PPM), they can support a 1 to 2 Mbps data rate.

The 802.11 standard can use either direct sequence or frequency hopping spread spectrum. Frequency hopping systems hop between 79 frequencies in the United States and Europe and 23 frequencies in Japan. Direct sequence achieves data rates of 2 Mbps, whereas frequency hopping can send data at 1 or 2 Mbps in the original 802.11 standard.

Because of the spatial expanse of wireless networks, the type of collision detection used in Ethernet would not work. Consider two stations, station 1 and station 2, that are not in range of each other. However, both are in range of station 3. Since the first two stations are not in range of each other, they could both transmit to station 3 simultaneously upon detecting an idle channel in their local geographic region. When both transmissions reach station 3, a collision results (i.e., overlapped garbled signals). This situation is called the hidden node problem (Tanenbaum 03).

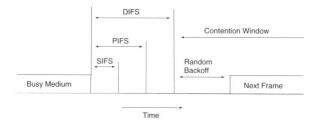

Fig. 1.12. Timing of 802.11 between two transmissions

To avoid this problem, instead of using CSMA/CD, 802.11 uses CDMA/CA (Carrier Sense Multiple Access with Collision Avoidance). To see how this works, consider only station 1 and station 3. Station 1 issues an RTS (request to send) message to station 3, which includes the source and destination addresses, the data type, and other information. Station 3, upon receiving the RTS and wishing to receive a communication from station 1, issues a CTS (clear to send) message signaling station 1 to transmit. In the context of the previous example, station 2 would hear the CTS and not transmit to station 3 while station 1 is transmitting. Note RTSs may still collide, but this would be handled by rescheduled transmissions.

The 802.11 protocol also supports asynchronous and time-critical traffic as well as power management to prolong battery life.

Figure 1.12 shows the timing of events in an 802.11 channel. Here after the medium becomes idle, a series of delays called spaces is used to set up a priority system among acknowledgments, time-critical traffic, and asynchronous traffic. An interframe space is an IFS.

Now, *after* the SIFS (short interframe space), acknowledgments can be transmitted. After the PIFS (point coordination interframe space), time-critical traffic can be transmitted. Finally, after the DIFS (distributed coordination interface space), asynchronous or data traffic can be sent. Thus, acknowledgments have the highest priority, time-critical traffic has the next highest priority, and asynchronous traffic has the lowest priority.

The 802.11 standard has an optional encryption protocol called WEP (wired equivalent privacy). A European competitor to 802.11 is HIPERLAN (and HIPERLAN2). Finally, note that wireless protocols are more complex than wired protocols for local area network and other environments.

Other 802.11 Versions

Since the original 802.11, a number of improved versions were developed and have become available. The original 802.11 version itself did not sell well as the price and performance was not that appealing. The three general variations are 802.11b, 802.11a, and 802.11g. Each is now briefly discussed. See Kapp 02 for a complete discussion. In 2004, almost 40 million WiFi units were shipped. More U.S. households now use WiFi for a home LAN rather than Ethernet.

802.11b: This 1999 version first made WiFi popular. It operates at a maximum of 11 Mbps at a range of 100–150 feet and 2 Mbps at 250–300 feet. Data rate decreases with distance to 1 Mbps and then goes to zero. If WEP is used with encryption, the actual useful data rate drops by 50%.

The 802.11b signal is in the 2.4 GHz band. It can operate either using direct sequence spread spectrum, frequency hopping, or infrared. Direct sequence is very popular, and infrared is mostly not in use.

802.11a Despite the name, 802.11a was developed after 802.11b. It operates at 54 Mbps in the UNI (Unlicensed Infrastructure Band):

Table 1.5. UNI Bands

Name	Band
UNI-1	5.2 GHz
UNI-2	5.7 GHz
UNI-3	5.8 GHz

There is some disagreement in the technical literature as to whether 802.11b or 802.11a has the larger range.

802.11g: Sometimes 802.11g is known as 802.11b extended. Initial versions were at 22 Mbps, later versions at 54 Mbps, and future versions will have higher data rates.

Note that some access points include options for more than one 802.11 version.

A number of specialized 802.11 standards have also been in development, as shown in Table 1.6.

Table 1.6. Other 802.11 protocols

Name	Description
802.11e	With quality of service
802.11h	Standard for power use and radiated power
802.11i	Uses WEP2 or AES for improved encryption
802.11x	Light weight version of EAP (Extended Authentication Protocol)

A word is in order on 802.11 security. A user requires some sophistication to prevent snooping by others. For instance, security features on shipped products are often disabled by default. Williams reports that many corporate users are not using or misusing WEP. There are media articles on people driving by in vans tapping into private networks. The 2001 article by Williams describes a series of security weaknesses in 802.11b. Some in the wireless LAN industry feel that, if one uses accepted security practices along with 802.11 features, security is acceptable.

1.6.2 802.15 Bluetooth

The original goal of Bluetooth technology, standardized as IEEE 802.15, is to provide an inexpensive, low power chip that can fit into any electronic device and use ad hoc radio networking to make the device part of a network. For instance, if your PC, printer, monitor, and speakers were Bluetooth enabled, most of the rat's nest of wiring under a desktop would be eliminated. Bluetooth chips could also be placed in PDAs, headsets, etc.

Work on Bluetooth started in 1997. Five initial corporate supporters (Erickson, Nokia, IBM, Toshiba, and Intel) grew to more than 1000 adopters by 2000. The name Bluetooth comes from the Viking King of Denmark, Harald Blatand, who unified Norway and Denmark in the tenth century (Haartsen 00).

Technically Speaking

Bluetooth had a number of design goals. As related in Haartsen 00, among them were:
- System should function globally
- Ad hoc networking
- Support for data and voice
- Inexpensive, minature, and low-power radio transceiver

Bluetooth operates in the 2.4 GHz ISM band (see the previous section's discussion of the ISM band). It uses frequency hopping spread spectrum (79 hopping channels, 1600 hops/second). Time is divided into 625 microsecond slots with one packet fitting in one slot. The data rate is 1 Mbps. The range is 10 meters, making Bluetooth a personal area network (PAN).

Two types of connections are possible with Bluetooth. First, SCO links (Synchronous Connection Oriented) are symmetrical, point-to-point, circuit-switched voice connections. Second, ACL links (Asynchronous Connectionless) are asymmetrical or symmetrical, point-to-multipoint, packet-switched connections for data.

A number of features of Bluetooth are designed to make possible good interference immunity. One is the use of high rate frequency hopping with short packets. There is an option to use error correction (see chapter 4) and a fast acting automatic repeat request scheme using error detection. Finally, voice encoding that is not that susceptible to bit errors is used.

Ad Hoc Networking

Two or more Bluetooth nodes form a "piconet" in sharing a frequency hopping channel. One node will become a "master" node to supervise networking. The other nodes are "slaves." Not only may roles change, but roles are lost when a connection is finished.

All SCO and ACL traffic is scheduled by the master node (Haartsen 00). The master node allocates capacity for SCO links by reserving slots. Polling is used by ACL links. That is, the master node prompts each slave node in turn to see whether it has data to transmit (see Schwartz 87 for a detailed discussion of polling). Slave node clocks are synchronized to the master node's clock.

There is a maximum of eight active nodes on a single piconet (others may be parked in an inactive state). As the number of nodes increases, throughput (i.e., useful information flow) decreases. To mitigate this problem, several piconets with independent but overlapping coverage can form a "scatternet." Each piconet that is part of a scatternet uses a separate pseudo-random frequency hopping sequence. The scatternet approach results in a very small decrease in throughput. Note that a node can be on several piconets in a scatternet. Moreover, it may be a master node on some piconets and a slave node on other piconets.

802.15.4 Zigbee

In actuality, the original Bluetooth has faced some problems in gaining acceptance. The rapid growth of 802.11 technology and its pricing has not given Bluetooth a price advantage on certain applications (Zheng 04). Also, Bluetooth is more complex than its original design goal as an attempt was made to have it serve more applications and supply quality of service. There is also some question on the scalability of scatternets.

Two successors to the original 802.15 Bluetooth standard are 802.15.3a for high-rate ultra wideband (UWB) wireless personal area networks (WPANs) and 802.15.4 for low-rate, low-power WPANs. In this section, the low data rate extension of Bluetooth is discussed.

A great many applications could benefit from a low data rate Bluetooth standard (Zheng 04). Among these are home, factory, and warehouse automation. These are applications for monitoring involving safety, the health field, and the environment. The use of low data rate Bluetooth for precision asset location and situation awareness could take place for emergency services and inventory tracking. Finally, there are potential entertainment applications including interactive toys and games.

Zigbee can operate either in the 2.4 GHz ISM band (available worldwide), the ISM 868 MHz band (Europe), and the ISM 915 MHz band (North America). Twenty-seven channels are defined for 802.15.4 as indicated in Table 1.7.

Zigbee, like Bluetooth, has a range of 10 meters. Communication can take place from a device, to a coordinator, a coordinator to a device, or between stations of the same type (i.e., peer to peer) in a multi-hop mode of operation. An 802.15.4 network can have up to 64,000 devices in terms of address

Table 1.7. Zigbee Channels

No. of Channels	Data Rate	Band
16 channels	250 kbps	2.4 GHz
10 channels	40 kbps	915 MHz
1 channel	20 kbps	868 MHz

space. Zigbee topology includes a one hop star or the use of multi-hopping for connectivity beyond 10 meters.

In beacon-enabled mode, the coordinator periodically broadcasts "beacons" to synchronize the devices it is connected to and for other functions. In non-beacon-enabled mode, beacons are not broadcast periodically by the coordinator. Rather, if a device requests beacons, the coordinator will transmit a beacon directly to the device (Zheng 04). A loss of beacons can be used to detect link or node failures.

It is critical for certain applications to minimize Zigbee coordinator and device energy usage. Some of these applications will be battery powered where batteries will not be (practically or economically) replaceable.

The majority of power-saving functions in 802.15.4 involve beacon-enabled mode. In direct data transmissions between coordinators and devices, the transceivers are only on 1/64 of the duration of a packetized superframe (i.e., collection of slots). A small CSMA/CD backoff duration, and brief warm-up times for transceivers, are also used to minimize power usage in 802.15.4.

Three security levels are available in 802.15.4. The lowest level is None Security mode, which is suitable if the upper layers provide security or security is not important. An access control list is used in the second level of security to allow only authorized devices to access data. The Advanced Encryption Standard (AES) is used in the highest, third security level.

1.6.3 802.16 Wireless MAN

Although wireless connectivity is very convenient, 802.11 and 802.15 have somewhat limited ranges (hundreds of feet and ten meters, respectively). A third, relatively recent wireless standard is IEEE 802.16 (Eklund 02, Tanenbaum 03). It defines how a base station may provide connectivity to computers in buildings up to several kilometers distant. A home or business owner may install an antenna that allows him or her broadband Internet connectivity while bypassing telephone or cable company wired services. Although current 802.16 technology uses fixed antennas in the subscriber's location, future technology may allow 802.16 connectivity to mobile users and directly to laptops.

Standard efforts on 802.16 began in 1999, and the standard was published in April 2001.

The original 802.16 operates in the 10–66 GHz band to make possible a large bandwidth and thus data rate. Precipitation can be a problem on this band so forward error correction (Reed Solomon) coding is used. There is also an option for the use of CRC coding (see chapter 4). Radio at 10–66 GHz is directional so base stations can have several antennas, each covering a different geographic sector.

The 802.16 standard calls for the use of either time division duplexing (TDD: base station and user antenna share a single channel divided into time slots but do not transmit concurrently) or frequency division duplexing (FDD: separate channels, sometimes transmit concurrently). Note that in TDD there is a variable upstream/downstream capacity allocation via time slot assignment.

These types of modulation are used in 802.16 depending on distance, as shown in Table 1.8.

Table 1.8. 802.16 Modulation Formats

Distance	Modulation	Bits per Symbol	No. of Waveforms
Small Distance	QAM-64	6 bits/symbol	64
Medium Distance	QAM-16	4 bits/symbol	16
Large Distance	QPSK	2 bits/symbol	4

In general, a digital modulation scheme encodes a symbol as one of a number of possible amplitudes and phase shifts. For instance, QAM-64 has 64 combinations of amplitude and phase shifts so the equivalent of 6 bits ($2^6 = 64$) can be transmitted per symbol. As distance increases, it is harder to distinguish between the 64 combinations due to noise and channel effects so that fewer, more distinctive combinations are used (16 for QAM-16 and then 4 for QPSK) with corresponding lower bit/symbol rate. Note that, if X MHz of spectrum is available, the data rates are 6X Mbps for QAM-64, 4X Mbps for QAM-16, and 2X Mbps for QPSK.

In terms of security, mutual authorization using RSA public key cryptography and X.509 certificates is specified in the standard.

A variety of downstream traffic can be supported in connection-oriented mode. The four supported classes are as follows:

- Constant bit rate.
- Real-time variable bit rate.
- Non-real-time variable bit rate.
- Best effort.

The physical and data link protocol layers for 802.16 appear in Figure 1.13.

The service-specific convergence sublayer interfaces to the network layer and is similar to the LLC (Logical Link Control) layer in older 802 standards. This includes interfacing for ATM (ATM convergence sublayer) and IPv4, IPv6, Ethernet, and VLAN (packet convergence sublayer). The channel is supervised by the MAC sublayer common part. Security (encryption, key management) is the responsibility of the security sublayer.

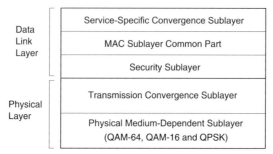

Fig. 1.13. Protocol stack for the 802.16 standard

Finally, there are standards spinoffs to implement 802.16 in the 2–11 GHz band (802.16a) and in the 5 GHz ISM band (802.16b).

1.7 ATM

The technology of ATM was developed in the telephone industry in the 1980s at the major telephone research laboratories. The goal was to develop a networking technology that could carry an integrated mixture of traffic (voice, video, and data). At one point, there were two competitors to be the technology of choice, synchronous transfer mode (STM) and asynchronous transfer mode (ATM). Asynchronous transfer mode was the eventual winner.

In fact, ATM is a packet-switched-based technology using virtual circuits. That is, all packets in a flow between two nodes follow the same path. Fixed length packets are used (53 bytes: 5 bytes of header and 48 bytes of data). Special architectures for high-speed, high-throughput, ATM packet switching have been developed and are discussed below.

Although a number of small companies attempted to market ATM local area networks, today ATM is found in the backbones of the Internet. It is not clear that it will have staying power. Some effort is underway to eliminate what some see as an unnecessary layer of complexity in protocol stacks.

Before discussing some aspects of ATM, let's first contrast it with STM.

1.7.1 Limitations of STM

Conceptually, STM (Littlewood 87) is closer to traditional telephone circuit switching than ATM. The basic idea is to build a high-speed channel out of a small set of basic channel building blocks. For instance, a need for a 200 Kbps channel is met by aggregating four standard 64 Kbps digital voice channels (called B channels). Or a need for 4 Mbps is met by aggregating three 1.536 Mbps H1 channels.

What were some problems with STM that led to its abandonment?

• The hierarchy is rigid. Note that, in the two examples above, because there are only a small number of basic channel data speeds, in most cases, some capacity is wasted once channels are aggregated.

• As in digital telephony, switching is done by time slot mapping. But STM leads to a complex time slot mapping problem.

• It was also felt that multi-rate switching was difficult to build out of 64 kbps building blocks.

• Separate switches may be needed for each type of traffic to be carried. This is a real show stopper. Traffic prediction, especially by class, is inexact and one could install many switches of some classes that could be under or over-utilized. In a national network, the economic loss could be very large.

• Circuit switching is not efficient for bursty (i.e., intermittent) data traffic (see Table 1.9) as capacity is allocated even if at that instant no data are transmitted. This is a waste of resources. It may not be practical to set up a channel only when bursty data are present due to the overhead of call setup time.

Table 1.9. Burstiness of Traffic

Class	Peak Rate (kbps)	Peak/Mean Bit Rate
Voice	16–64	2:1
Text	1–64	2:1
Image/Data	64–2000	10:1
Video	$\leq 140{,}000$	5:1

Packet switching and ATM, on the other hand, are particularly suited for bursty traffic. There is what is called statistical multiplexing and inherent rate adaptation. That is, links and other resources are efficiently shared by multiple source/destination pairs. If some sources are idle, other busy sources can easily make use of link capacity.

Moreover, there is no need for different types of switches for each service class. From the viewpoint of the 1980s, one could justify the investment in a single type of switch on the majority of voice traffic (that has since changed) and experimentally run new services on top of that at relatively low marginal cost. In fact, in a sense the design problem for a single type of switch is simply to design the highest throughput switch for the amount of money available. In fact, things are really a bit more complex (there's quality of service, for instance), but the design problem is still simpler compared with the STM alternative.

1.7.2 ATM Features

Once again the ATM packet (called a "cell" in ATM language) is 53 bytes long, 5 bytes of which are header. There are actually two types of headers,

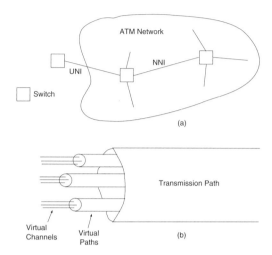

Fig. 1.14. An (a) ATM network and (b) a transmission path broken into virtual paths and virtual channels

depending on whether the packet is traversing a link at the boundary of the ATM network (UNI: user-to-network interface) or a link internal to the ATM network (NNI: network-to-network interface). See Figure 1.14(a) for an illustration (Onvural 95). In ATM language, a generic virtual circuit is called a "virtual channel." Virtual channels are bundled into "virtual paths." A link carries multiple virtual paths, each consisting of many virtual channels [see Figure 1.14(b)].

Let's examine the UNI header first (Figure 1.15). The first 4 bits are for generic flow control (GFC), a field that was incorporated in the header but in reality is not used. There are 8 bits for virtual paths (thus, $2^8 = 256$ virtual paths/link) and 16 bits for virtual channels (thus, $2^{16} \cong 64,000$ virtual channels/virtual path). The 3 bit payload type (PT) field indicates eight possible payload types. The 1 bit cell loss (CL) priority field indicates whether the cell can be dropped under congestion. Finally, the header error check field uses a code to protect the header only, with single bit error correction and detection for 90% of multiple bit errors. Note that, if error protection is needed for the data, this has to be taken care of at a different protocol layer.

The NNI header fields are similar to the UNI fields except that there is no generic flow control field in the NNI header and 12, rather than 8, bits are reserved for virtual paths. Thus, the NNI has $2^{12} = 4096$ virtual paths compared with the $2^8 = 256$ virtual paths for the UNI. This is because the NNI links are more like internal trunks, which carry more traffic than the access links, as with the UNI. Note that, since the number of bits in the virtual channel field is the same for both the UNI and the NNI, a single virtual path contains the same number of virtual channels (2^{16}) whether it involves the UNI or NNI.

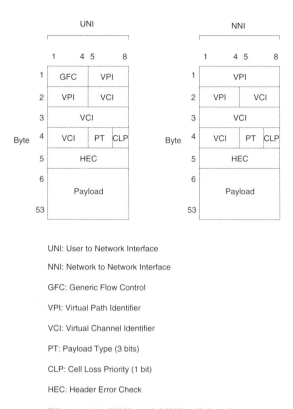

UNI: User to Network Interface

NNI: Network to Network Interface

GFC: Generic Flow Control

VPI: Virtual Path Identifier

VCI: Virtual Channel Identifier

PT: Payload Type (3 bits)

CLP: Cell Loss Priority (1 bit)

HEC: Header Error Check

Fig. 1.15. UNI and NNI cell headers

The 53 byte packet size for ATM was chosen as a compromise among interested parties and partly serves to minimize queueing delay through the use of a relatively short packet.

A word is in order on the payload type field. The eight (2^3) possibilities are (Tanenbaum 03) as follows in Table 1.10.

Table 1.10. Payload Types

Type	Explanation
000	User data cell, no congestion, packet type 0
001	User data cell, no congestion, packet type 1
010	User data cell, congestion present, packet type 0
011	User data cell, congestion present, packet type 1
100	Management info for adjacent ATM switches
101	Management info for source/destination ATM switches
110	Resource management cell
111	Reserved for future use

It is a bit redundant, but cell type (whether a packet can be dropped under congestion) can be indicated through either the payload type or the cell loss priority field.

Special resource management (RM) cells are inserted periodically on virtual channel streams for congestion control and then returned to the source. An RM cell that does not return to the source in a reasonable time indicates congestion. An explicit transmission rate in the RM cell can also be lowered by congested switches to throttle back the source. Overloaded switches can also create RM cells. This rate-based congestion control is used in ATM, but a good discussion of other discarded possibilities for congestion control appears in the third edition of Tannenbaum 96.

In order to carry a mixture of traffic, ATM supports four classes of traffic. They are as follows:

- Constant bit rate (CBR).
- Variable bit rate (VBR), which consists of the real-time (RT-VBR) and the non-real time (NRT-VBR) option.
- Available bit rate (ABR).
- Unspecified bit rate (UBR).

The available bit rate class may guarantee a minimum data rate but exceeds that sometimes. The unspecified bit rate class gives no guarantees and can be used for file transfers or email.

ATM technology can provide quality-of-service (QoS) guarantees. Two communicating nodes agree (through a "contract") on QoS parameters specifications. There are a large number of QoS parameters, such as minimum and peak cell rate, cell transfer delay, and cell error rate.

1.7.3 ATM Switching

A "switch" is a computerized device that interconnects users or computers. In other words, a switch provides connectivity. Switching can in general be based on either circuit switching or packet switching. There are three general architectures for building ATM packet switches: shared medium, shared memory, and space division switching (Robertazzi 03). Each is now discussed in turn.

Shared Medium Architecture

This architecture uses a computer bus to provide the interconnection capability. An illustration appears in Figure 1.16. A computer bus is a fairly large number of parallel wires. Although each operates at a moderate speed, the aggregate rate is impressive. For instance, a 64 bit bus with each wire at 80 Mbps has an aggregate data rate in the gigabits per second (i.e., $64 \times 80 \times 10^6$ bps).

Several buffers provide input and output paths to the bus. The buffers exiting the bus are shown larger than the input buffers as it is usual for the

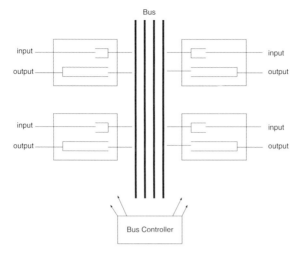

Fig. 1.16. Shared medium switch architecture

aggregate speed of the bus to be significantly larger than the input port rate. A bus controller coordinates access to the bus.

A little thought will show that, for a switch with N inputs, the bus should operate at least N times faster than the input packet arrival rate. Thus, the switch is said to have an N times speed-up.

Shared Memory Architecture

A shared memory switch architecture is shown in Figure 1.17. The inputs are multiplexed into a single stream accessing a dual port memory. The memory is organized into packet buffers and address chain pointers.

The memory is partitioned by the output ports to which the packets are addressed. At one extreme is "full sharing" where the complete memory is shared by all output ports. A problem with full sharing is that a heavily

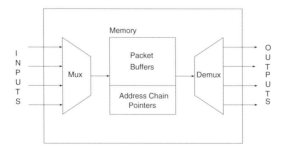

Fig. 1.17. Shared memory switch architecture

used output port's packets may occupy most or all of the memory, leading to a problem of fairness for other ports. One could implement "complete partitioning" to address the problem. Under complete partitioning, $1/N$ of the memory is dedicated exclusively to each output port. Although this solves the original problem, the downside is that, when a partition fills up, space may be available in other partitions that cannot be used.

Packets from the memory leave the other memory port, are demultiplexed, and sent to their respective output ports.

Note that the memory access speed must be at least N times faster than the individual packet arrival rate for each input.

Space Division Switches

Space division switches use several parallel paths in the switch to boost throughput (i.e., traffic flow). Space division switches are usually built using patterned networks of small switching elements. The fact that the same basic switching element is replicated many times makes this type of design suitable for VLSI implementation. There are many types of space division switches. Some particular examples, such as crossbars, Banyan networks, and the knockout switch are discussed in chapters 2 and 4.

1.8 SONET

SONET is a popular standard for fiber-optic voice and data transmission. It was developed originally by Bellcore, the research and development arm of the local American phone companies in the late 1980s (Siller 96). It was meant to be a standard for fiber-optic connections between telephone switches. However, it was a technology at the right place, at the right time, and has been extensively used over the years for telephone trunk transmission and internal corporate and governmental traffic. More specifically, it was developed at about the time that there was an interest in providing broadband integrated services digital network (B-ISDN) technology.

SONET, when it was developed, took into account B-ISDN technology, political, and international compatibility concerns. The SONET architecture is elegant and took advantage of LSI and software advances at the time. Development has continued over the years with the introduction of higher and higher standardized data rates.

A typical SONET data rate is abbreviated as STS-n/OC-n, where $n = 1, 2, 3 \ldots$. The "STS" indicates the electrical interface, and the "OC" indicates the optcal interface. The STS-1/OC-1 rate is 51.84 Mbps. Any other STS-n/OC-n rate is n times faster than 51.84 Mbps. For instance, STS-3/OC-3 is at 155.52 Mbps. In fact, STS-3/OC-3 is the lowest SONET rate used in practice. Table 1.11 indicates some of the various SONET rates.

Table 1.11. Some SONET Rates

Acronym	Gross Rate
STS-1/OC-1	51.84 Mbps
STS-3/OC-3	155.52 Mbps
STS-12/OC-12	622.08 Mbps
STS-48/OC-48	2.48832 Gbps
STS-192/OC-192	9.95328 Gbps
STS-768/OC-768	39.81312 Gbps

Lower rates, known as virtual tributaries, are also available. For instance, virtual tributary 1.5 (VT1.5) is at 1.728 Mbps. Some virtual tributary rates are indicated below.

Table 1.12. Virtual Tributary Rates

Acronym	Data Rate
VT1.5	1.728 Mbps
VT2	2.304 Mbps
VT3	3.456 Mbps
VT6	6.912 Mbps

Note that VT1.5 is compatible with the T1 rate of 1.544 Mbps and VT2 is compatible with the European version of the T1 rate of approximately 2.0 Mbps. The European version of SONET is SDH (Synchronous Digital Hierarchy).

1.8.1 SONET Architecture

SONET data are organized into tables. For STS-1/OC-1, the byte table consists of 9 rows of bytes and 90 columns of bytes (Figure 1.18). As shown in the figure, the first 3 columns hold frame overhead and the remaining 87 columns hold the payload. Some additional overhead may appear in the payload. Each byte entry in the table holds 8 bits. If digital voice is being carried, the 8 bits represent one voice sample. Uncompressed digital voice consists of 8000 samples/second of 8 bits each (or 64 Kbps). Thus, the SONET STS-1/OC-1 frames are generated at a rate of 8000 frames/second.

The protocol layers for SONET go by the names of path, line, section, and photonic (see Figure 1.19 where ATM is being carried over SONET). The functions of the layers are (Onvural 95):

• Path = End-to-end transport as well as functions including multiplexing and scrambling.

• Line = Functions include framing, multiplexing, and synchronization.

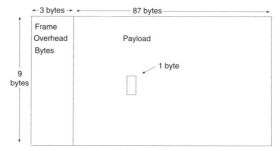

Fig. 1.18. An STS-1/OC-1 SONET frame

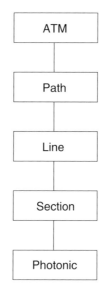

Fig. 1.19. SONET protocol stack

- Section = Functions include bit timing and coding.
- Photonic = Physical medium.

Overhead of each type appears in an STS-1/OC-1 frame as illustrated in Figure 1.20.

Note that the start of a payload is indicated by a pointer in the line overhead.

The protocol layers can be viewed in terms of a box type diagram as in Figure 1.21.

There are two major configurations for the SONET opto-electronic interface at a node. If the fiber starts/ends at the node, one says one has a SONET Add Drop Multiplexer (ADM) in terminal mode. The ADM allows signals to be tapped off or on the fiber. Alternatively, one may have fiber passing through the node. That is, a fiber enters from the east, is converted to an electrical

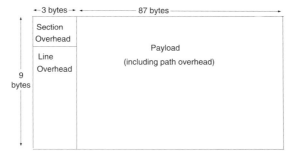

Fig. 1.20. An STS-1/OC-1 SONET frame with overhead indicated

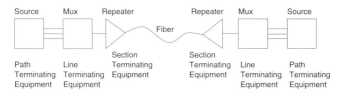

Fig. 1.21. SONET system diagram

signal, signals are tapped off and inserted, and a new fiber leaves to the west. This is called a SONET ADM in add/drop mode.

1.8.2 Self-Healing Rings

One reason for the widespread use of SONET is it allows a variety of topologies. The most useful of these are ring topologies. Although one can implement linear add/drop networks (see Figure 1.22), if a fiber is cut or an optoelectronic transceiver fails, one loses complete connectivity.

Typically, rings are implemented with multiple service and protection fibers (see Figure 1.23). If a service fiber path fails, a protection (back-up) fiber is switched on in its place. The number of protection fibers can be less than the number of service fibers if one is willing to live with less redundancy. Also, if all fibers between two adjacent nodes are lost, with a sufficient number of protection fiber, rerouting can keep a logical ring in place until repairs are made.

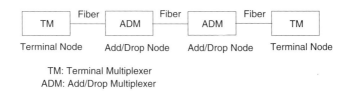

Fig. 1.22. A SONET linear network

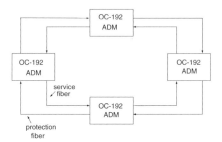

Fig. 1.23. A SONET ring network

SONET rates have increased over the years but not by enough for fiber to reach its full potential unless a second technology, wavelength division multiplexing, is also used. This WDM technology is discussed in the next section.

1.9 Wavelength Division Multiplexing (WDM)

In WDM, special multiplexing at either end of a fiber can put multiple optical signals in parallel on a single fiber (Ryan 98). Thus, instead of carrying one OC-192 signal at about 10 Gbps, a single fiber might carry 40 OC-192 signals or 400 Gbps of traffic. Even terabit range (1000 Gbps) capacity is possible. With each signal at a different optical frequency, WDM is essentially a frequency division multiplexing scheme.

The history of WDM goes back to a fiber glut in the United States that existed prior to 1995. After the Bell System divestiture in the 1980s, the competitors in long-distance phone service had financial limitations so relatively low numbers of fiber per path were laid (usually about 16 fibers). But by the end of 1995, the interexchange fibers of the long-distance carriers were nearing exhaust. In 1996, 60% of the AT&T network, 84% of the MCI network, and 83% of the Sprint network were fully lit (Ryan 98).

About this time WDM technology became practical. This technology included distributed feedback lasers needed to produce the monochromatic output of WDM, filters to separate signals that are closely packed in frequency, and optical amplifiers. In particular, Erbium Dopel Fiber Amplifiers (ED-FAs) allow the amplification of optical signals without intermediate electronic conversion.

In 1994, Pirelli/Nortel introduced 4 channel systems and IBM introduced a 20-channel system. Cienna followed with a 16-channel system in 1996. By 1997–1998, 32s and 40-channel systems were being produced. It should be noted that Cienna's successful WDM products led to a very successful public offering and $196 million in first-year revenue (the fastest in corporate history at that point).

Now, using WDM in conjunction with SONET, if one had 32 OC-48 channels, rather than using 32 separate fibers, one could use a 32λ system (32λ means 32 wavelengths) on a single fiber or 8 fibers each with OC-192.

At first, WDM technology was used in long-distance networks, but as its costs decreased, metropolitan area network usage followed.

Tunable lasers have been introduced as a way of providing back-up. That is, rather than having 176 fixed wavelength spares for a 176λ system, one tunable laser provides protection against the most likely case, a single bad fixed wavelength laser.

1.10 Grids

A grid is a distributed computing system that allows users to access large amounts of computer power and data storage over a network to run substantial computing tasks. Ian Foster, a leader in grid development, has written (Garritano) that a grid must
- Provide resource coordination without a central control.
- Use standardized and open protocols and interfaces.
- Provide substantial amounts of service involving multiple resource types and nontrivial user needs.

As Schopf 02 points out, the idea of applying multiple distributed resources to work on a computational problem is an old one. It goes back at least to the mid-1960s and the "computer utility" paradigm of the MULTICS operating system and work on networked operating systems. Additional work on distributed operating systems, heterogeneous computing, parallel distributed computing, and meta-computing further explored this area.

Work on the grid concept started in the mid-1990s. Among significant grid features that make it a distinctive problem area are:
- Grids allow site autonomy. Local sites have control over their own resources.
- Grids allow heterogeneity. Grid development work provides standardized interfaces to overcome site diversity.
- Grids are about data (representation, replication, storage, in addition to the usual network and computer issues).
- The key person for grids is the user, not the resource owner. Earlier systems sought to boost utilization/throughput for the resource owner. In a grid, machines are selected to fulfill the user's requirements.

A great deal of grid work is related to interfaces and software development to allow different sites to work in concert. However, the grid effort has taken time, enough time that some have questioned its practicality. A balanced discussion of the difficulties and challenges facing grid development appears in Schopf and Nitzberg. Here we mention some of these problems.
- In many cases, grids are being used to solve for embarrassingly parallel applications rather than for coordinated distributed computing.

- Simple interoperability is still to come. Users often have to go through a great deal of work to achieve even basic functionality.
- Some science users see making applications suitable for the grid as a "distraction" from getting the science accomplished.
- It is difficult for developers to know what tools to develop when there is no user experience because there are no tools. Applied scientists don't know what to request from developers.
- Funding for adapting applications to a grid environment can be drained, if not blocked, by installation and administration problems.
- System administrators are used to having control over their local resources so grid software can be seen as "threatening."
- Setting up an account in a distributed grid can be complex.
- The recent adoption of web services has led to competing incompatible grid software incarnations, leading to a difficult situation for grid developers in terms of deciding which software will have staying power.

All of this is not to say that the vision of successful grids makes overcoming these difficulties and growing pains not worthwhile.

The Open Grid Service Architecture is an important grid standard announced in 2002. In 2004, OGSA version 1.0 was released. Version 2.0 is expected to be released in 2005. The service oriented architecture of OGSA is illustrated in Figure 1.24.

In the figure, note that OGSA services are presented to the application layer. Also, OGSA services make use of web services. Note also this grid architecture has three hardware components: servers, storage, and networks. A discussion of OGSA instantiations appears in Baker 05.

The main grid standards setting body is the Global Grid Forum. Other organizations with grid involvement include the Organization for the Advancement of Structured Information Standards (OASIS), the World Wide

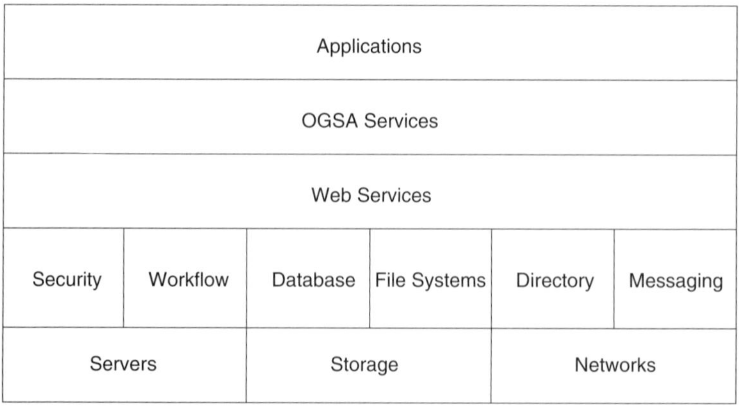

Applications					
OGSA Services					
Web Services					
Security	Workflow	Database	File Systems	Directory	Messaging
Servers		Storage		Networks	

Fig. 1.24. Open Grid Service (OGSA) Architecture

Web Consortium, the Distributed Management Task Force, the Web Services Inter-operability Organization, and some groups in Internet 2 such as the Peer-to-Peer Working Group, the Middleware Architecture Committee for Education, and the Liberty Alliance (Baker 05).

For further information, readers are referred to Foster 03, Baker 05, and www.globus.org.

1.11 Problems

1. Describe one advantage of using microwave line-of-sight technology compared with wired technologies.
2. Which type of communication link technology has the largest information carrying capacity?
3. What is the "electronic bottleneck" associated with fiber optics?
4. Name and explain a clear advantage of geosynchronous satellite technology over LEOS technology. How about the reverse?
5. Name and describe an alternative to wires, fiber, cables, or radio waves for connecting a computer to a network.
6. How does ad hoc radio transmission differ from satellite transmission?
7. Name a potential application of wireless sensor networks.
8. Why can the header of a packet be larger at the physical layer than at a higher layer such as the network or transport layers?
9. Why do packets transmitted over a virtual circuit arrive in the same order they were transmitted?
10. In the OSI reference model, a network layer entity at one node wishes to communicate with a network layer entity at another node. Describe the path the communications takes.
11. Name and explain a function of the data link layer.
12. What is the difference between the data link and network layers?
13. Which layer of the OSI protocol stack is responsible for providing end-to-end communications over possibly unreliable sub-networks?
14. Why does throughput decrease for Ethernet under heavy load?
15. Explain why longer frames lead to a higher utilization than short frames in Ethernet.
16. Suppose that you are designing a LAN with two signal paths from the hub to each PC. Ternary signaling is used. How many equivalent bits can be sent at once? Use the nearest power of two.
17. Why is an unshielded twisted pair version of Gigabit Ethernet a good thing?
18. In a wireless LAN, node A issues an RTS but does not receive a CTS from node B. What might A reasonably do to contact B?
19. Suppose node A is transmitting to node B using the IEEE 802.11 wireless LAN protocol. Node A sends an RTS to node B. Node B then transmits a CTS after which node A can send its data to node B. Why does the

protocol use a CTS on the part of node B? Could A transmit directly to node B without a CTS if it senses the channel to be idle?

20. Why do the various 802.11 versions operate in the Mbps range while fiber has an information carrying capacity in the Gbps range?

21. An auditing team travels from company to company and needs to set up a local area network at each company they audit. Which technology would be easier to set up and take down when they are finished: some version of Ethernet or some combination of 802.11 and 802.15? Explain.

22. Under STM (an old circuit-switched design competitor to ATM) if one needed an 80 kbps channel, one would allocate two 64 kbps B channels to construct a circuit with sufficient capacity. What is a major problem with this approach in terms of the efficient allocation of bandwidth?

23. In an NNI connection in an ATM network, what is the maximum number of virtual channels available?

24. Why does the header error check field in an ATM packet protect only the header and not the data?

25. What feature of the ATM packet makes misrouting of a packet extremely rare?

26. Why does the fact that ATM technology leads to service class independent switches a good thing?

27. Cell transfer delay is an ATM quality-of-service parameter that measures how long cell (packet) delivery takes. Name one application where it would be critical to get cells delivered very quickly.

28. Why is a "contract" for an ATM session difficult to define?

29. Compare the data rates of a T1 line and a SONET OC-3 channel.

30. Approximately how many STS-1/OC-1 frames are needed to transport 1 Mb? Neglect path overhead.

31. What is the function of an ADM in SONET?

32. What is the purpose of protection fibers in SONET?

33. Which SONET layer is most similar to the data link layer in the OSI reference model?

34. Why are more virtual paths allowed in an ATM NNI than in a UNI?

35. What is the effective data rate available in transmitting in the same byte entry in consecutive SONET OC-1 frames? What is the significance of this data rate?

36. What is the approximate data rate of OC-3072?

37. Approximately how many WDM channels are needed to carry 760 Gbps if each channel is OC-192?

38. What is the economic incentive for installing WDM systems? Say more than just for large bandwidth.

39. What is the basic idea of grid computing?

40. How do the security needs of an individual computer installation affect the installation's participation in grid computing?

41. Comment on the fact that the key person in a grid is the user, not the resource owner. How does this affect grid implementation efforts?

2

Fundamental Stochastic Models

2.1 Introduction

A "switch" (or "hub") is a computerized device that provides electronic connectivity among PCs, workstations, wireless devices, and other computes connected to it. This naturally means a switch will have a number of input and output links (connections) to these various user devices (Figure 2.1).

A packet switch accepts packets of data on incoming links and routes them on outgoing links toward their destination. Note that a packet (i.e., bundle of bits) typically consists of a header and the payload. The header holds control information such as source and destination addresses, packet priority level, and error checking codes. The payload is the actual information to be transmitted. Such a packet switching service is also known as connectionless or datagram service. Packet switching service has been of much interest since the implementation of early wide area packet switching networks or datagram service, starting in the 1960s.

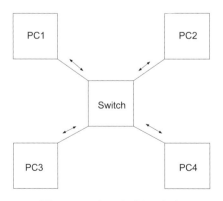

Fig. 2.1. A switching hub

An older technology compared with packet switching is circuit switching. Here traditionally a physical path is reserved from one local switch to a (possibly distant) switch. Thus, in the classic telephone network, the path extends from one phone to another passing through a number of intermediate (hierarchical) switches on the way. The use of circuit switching sometimes is said to provide connection-oriented service.

In both cases, the fundamental design problem is how does one design and build the highest performance switch. Performance typically involves such parameters as the best values of throughput (capacity) and time delay (to transmit a packet or setup a call) for the money invested. Typical approaches for designing the internal architecture of a high-speed packet switch could include the use of a computer bus, shared memory, or very large-scale integration (VLSI) switch designs based on space division switching.

A major goal of a performance evaluation study is to determine the performance of various architectural alternatives using statistical and mathematical models. Although experimental work is crucial in engineering a system that is reliable and has high performance, performance evaluation allows a cost-effective, preliminary consideration of technological alternatives prior to implementation. Furthermore, analytical performance evaluation results can give an intuitive understanding of design trade-offs. The rest of this chapter will discuss two paradigms for modeling arriving streams of traffic. These are the continuous time and discrete time paradigms. Probabilistic models are emphasized in this chapter because of their simplicity and utility.

In section 2.2, the fundamental Bernoulli and Poisson processes are discussed. Bernoulli process statistics are presented in section 2.3. Multiple access performance is covered in section 2.4, which includes discussions of Ethernet models and Aloha communication. Teletraffic modeling for specific topologies (linear, tree, and circular area topologies) appears in section 2.5. Finally switching elements and fabrics are examined in section 2.6.

2.2 Bernoulli and Poisson Processes

Discrete time systems typically involve the use of equi-length time slots. That is, if one plots the signal flow on a single input link versus time, one might have the situation at Figure 2.2a.

This figure is based on the Bernoulli process assumption that the probability of a (single) packet being present in a particular time slot is p. The probability of no packet being present in a particular slot is $1 - p$. Here each event in a slot (a packet or no packet) is independent of all others (which will simplify calculations below).

There are possible variations of the Bernoulli process, including allowing 0 to N arrivals, per slot. This will occur, for instance, in modeling a switching element (module) in a VLSI switching array with N inputs that are each

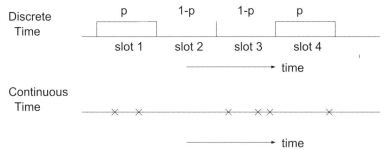

Fig. 2.2. Bernoulli and Poisson processes

mathematically modeled as a Bernoulli process input. Naturally we are assuming here that time slot boundaries are aligned so that in each slot 0 to N packets arrive simultaneously to the switching element.

Equi-length time slots are used in a high-speed packet switching technology known as ATM (Asynchronous Transfer Mode). ATM "cells" (really packets) are always 53 bytes or 424 bits long. ATM is widely used in Internet high-capacity backbones.

In some systems, a number of slots comprising a "frame" (say slot 1 through 100) repeat in a periodic fashion so that each user gets one slot of transmission time every time the frame repeats. Time division multiplexing, for instance, works in this manner.

In order to examine the Bernoulli process, let

$$\text{Prob}(1 \text{ arrival/slot}) = P_1 = p \tag{2.1}$$

$$\text{Prob}(\text{no arrival/slot}) = P_0 = 1 - p \tag{2.2}$$

$$\text{Prob}(2 \text{ or more arrivals/slot}) = 0.0 \tag{2.3}$$

One can then compute statistics for the Bernoulli process such as the mean (average) number of customers per slot or the variance in the number of customers per slot.

So, from first principles, the mean or expected number of packets $\bar{n}$ is

$$E[n] = \bar{n} = \sum_{i=0}^{1} iP_i = 0 \times P_0 + 1 \times P_1 \tag{2.4}$$

$$\boxed{\bar{n} = P_1 = p} \tag{2.5}$$

This result is quite intuitive. If there were always no arrivals, one would anticipate $E[n] = 0$. If each and every slot carries a packet, one would expect

$E[n] = 1$. In fact the probability p is between 0 and 1. One might consider p as a performance measure called "throughput." Throughput is the (possibly normalized) amount of useful information carried per unit time by a network element, be it a link, buffer, or switch.

Next, from first principles, one can compute the variance of the number of packets on a link as

$$\sigma^2 = \sum_{i=0}^{1} (i - E[n])^2 P_i \tag{2.6}$$

$$\sigma^2 = (E[n])^2 P_0 + (1 - E[n])^2 P_1 \tag{2.7}$$

$$\sigma^2 = p^2(1 - p) + (1 - p)^2 p \tag{2.8}$$

$$\sigma^2 = p(1 - p)[p + 1 - p] \tag{2.9}$$

$$\boxed{\sigma^2 = p(1 - p)} \tag{2.10}$$

Recall that variance is the sum of the squared differences between the values of a random variable and the random variable's mean, which are weighted by the probability of the random variable taking on specific values. Note that the variance in the number of customers per slot (above) is maximized when p is close to 0.5.

At this point, it should be noted that popular transform techniques (such as Laplace and Fourier transforms) allow one to treat a time-based signal in the frequency domain, and for certain types of work, this simplifies calculations. In fact a random process can often be transformed into a frequency-like domain using the moment generating function (really a z transform using z^n rather than signal processing's z transform's usual z^{-n}). From the moment-generating function definition, one has for the Bernoulli process

$$P(z) = \sum_{i=0}^{1} P_i z^i \tag{2.11}$$

$$P(z) = P_0 z^0 + P_1 z^1 \tag{2.12}$$

$$P(z) = (1 - p)z^0 + pz \tag{2.13}$$

$$\boxed{P(z) = (1 - p) + pz} \tag{2.14}$$

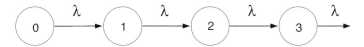

Fig. 2.3. Poisson process state transition diagram

Although no further use of the moment-generating function appears in this text, it is a useful analytical tool for such purposes as finding transient and advanced queue discipline statistics.

The continuous time analog to the discrete time Bernoulli process is the Poisson process. With continuous time modeling, an event, such as a switch arrival, may occur at any time instant (e.g., $t = 2.49346$ seconds), rather than at integer slot times (slot 1, slot 2...). It can be seen from Figure 2.2b that arrivals are random in placement. If one takes a certain interval of a homogeneous (i.e., constant arrival rate) Poisson process with M arrivals, the M events are in fact randomly placed in a uniform manner over the interval. Although the time-invariant or homogeneous Poisson process is quite tractable, time-varying arrival rates $[\lambda(t)]$ can also be modeled.

To find the basic equations governing a Poisson process, let $P_n(t)$ be the probability of n arrivals in a time interval of duration t.

As will be discussed, a state machine-like diagram called a state transition diagram can be drawn. The circles (Figure 2.3) are states $(0, 1, 2, \ldots$ customers having arrived up to this time), and transitions between them are labeled with the rate of making a transition from state i to state $i + 1$.

In fact we will initially use a time-slotted model to characterize the Poisson process. However these are not the fixed engineered macroscopic packet transmission times of the Bernoulli process. Instead the "slots" are the usual mathematically arbitrarily small intervals from $t \to t + \Delta t$ where, we'll eventually let $\Delta t \to 0.0$ to create a continuous time model.

Going through the steps, from first principles one has the difference equations $n = 1, 2 \ldots$

$$P_n(t + \Delta t) = P_n(t)P_{n,n}(\Delta t) + P_{n-1}(t)P_{n-1,n}(\Delta t) \tag{2.15}$$

$$P_0(t + \Delta t) = P_0(t)P_{0,0}(\Delta t) \tag{2.16}$$

Here $P_{n,n}(\Delta t)$ is the probability of going from n arrivals at a point in time $[t]$ to n arrivals at time $[t, t + \Delta t]$. Intuitively, $P_{n-1,n}(\Delta t)$ should be proportional to the arrival rate and the time slot width. Thus as Δt becomes small

$$P_{n-1,n}(\Delta t) = \lambda \Delta t \tag{2.17}$$

$$P_{n,n}(\Delta t) = 1 - \lambda \Delta t \tag{2.18}$$

Substituting, using algebra and letting $\Delta t \to 0$, one arrives at

$$\frac{dP_n(t)}{dt} = -\lambda P_n(t) + \lambda P_{n-1}(t) \quad n = 1, 2 \ldots \tag{2.19}$$

$$\frac{dP_0(t)}{dt} = -\lambda P_0(t) \tag{2.20}$$

Note that for n arrivals we have a family of n linear differential equations. The second equation electrical and computer engineers will recognize as being similar to the differential equation for capacitive voltage discharge through a resistor.

It is well known that the solution to this equation is

$$P_o(t) = e^{-\lambda t} \tag{2.21}$$

Substituting this solution into the $n = 1$ equation and solving will yield

$$\frac{dP_1(t)}{dt} = -\lambda P_1(t) + \lambda e^{-\lambda t} \tag{2.22}$$

$$P_1(t) = \lambda t e^{-\lambda t} \tag{2.23}$$

Continuing to substitute the current solutions into the next differential equation yields

$$P_2(t) = \frac{\lambda^2 t^2}{2} e^{-\lambda t} \tag{2.24}$$

$$P_3(t) = \frac{\lambda^3 t^3}{6} e^{-\lambda t} \tag{2.25}$$

Or simply

$$\boxed{P_n(t) = \frac{(\lambda t)^n}{n!} e^{-\lambda t} \quad n = 0, 1, 2 \ldots} \tag{2.26}$$

This is the Poisson distribution. Given the Poisson arrival rate λ and time interval length t, one can use the Poisson distribution to easily find the probability of there being n arrivals in the interval. Figure 2.4 illustrates the Poisson distribution for $n = 0$ to $n = 5$.

In fact it is straightforward to find the average or mean number of arrivals in a time interval of length t. From first principles (see the appendix), one has

$$\bar{n} = \sum_{n=0}^{\infty} n P_n(t) \tag{2.27}$$

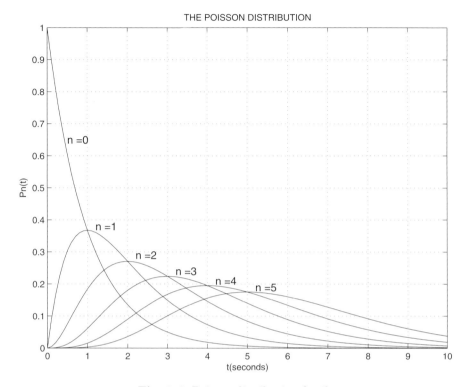

Fig. 2.4. Poisson distribution family

Substituting the Poisson distribution, one has

$$\overline{n} = \sum_{n=0}^{\infty} n \frac{(\lambda t)^n}{n!} e^{-\lambda t} \tag{2.28}$$

As $e^{-\lambda t}$ is not affected by the summation index, n, one can move $e^{-\lambda t}$ past the summation. Also clearly the $n = 0$ term is zero as n is a factor. Thus

$$\overline{n} = e^{-\lambda t} \sum_{n=1}^{\infty} \frac{(\lambda t)^n}{(n-1)!} \tag{2.29}$$

With a change of variables $(n - 1) \rightarrow n$, one has

$$\overline{n} = e^{-\lambda t} \sum_{n=0}^{\infty} \frac{(\lambda t)^{n+1}}{n!} \tag{2.30}$$

Again a term, this time a factor of λt, doesn't depend on n and so the factor can be brought outside the summation

$$\bar{n} = e^{-\lambda t} \lambda t \sum_{n=0}^{\infty} \frac{(\lambda t)^n}{n!} \tag{2.31}$$

Finally using a summation formula from the appendix, one has

$$\boxed{\bar{n} = e^{-\lambda t}(\lambda t)e^{\lambda t} = \lambda t} \tag{2.32}$$

The fact that the average number of customers arriving in an interval is λt is quite intuitive. For instance, if 100 calls arrive per minute to a telephone switch, the average number arriving in 10 minutes is $10 \times 100 = 1000$ calls. In fact it can be seen in Figure 2.4 where $\lambda = 1$ that the distributions $n = 1, 2, \dots$ peak at λt.

It is useful to know that, if N Poisson processes, each with Poisson arrival rate λ_i, are merged, the merged process is Poisson with mean arrival rate $\sum_{i=1}^{N} \lambda_i$. Also, suppose a Poisson process of rate λ is split randomly into N processes. That is, an arrival is sent to split process i with independent probability p_i. Naturally

$$p_1 + p_2 + p_3 \dots + p_N = 1$$

Then each split process is an independent Poisson process with rate $p_i\lambda$. Naturally merged processes arise in multiplexing, and split processes arise in demultiplexing.

Interestingly, the homogeneous Poisson process can be used as a model of the spatial (geographic) distribution of station location in two dimensions (Diggle 83). In this case λ_A is the average number of stations located in a unit area. Within an actual area A of any shape, the actual number of stations placed is uniformly (perfectly randomly) distributed.

Using the one-dimensional Poisson distribution as a starting point, one has the probability that there are n stations in area A, with spatial location density λ_A as $P_n(A)$

$$P_n(A) = \frac{(\lambda_A A)^n}{n!} e^{-\lambda_A A} \tag{2.33}$$

Intuitively the average number of station placed in area A is

$$\bar{n} = \lambda_A A \tag{2.34}$$

Other models of spatial location are possible (Miller 02, Cheng 90a).

2.3 Bernoulli Process Statistics

Given a discrete time Bernoulli process, a number of standard distributions are associated with natural questions about the process. Three such distributions are discussed below.

Question 1: How many slots does it take, on average, for the first arrival?

To answer this question, use can be made of the geometric distribution. Consider a data record of $i - 1$ idle slots followed by the ith slot containing a packet. Naturally p here is the probability of an arrival in a slot. The probability of no arrival in a slot is $1 - p$. It is assumed that arrivals/nonarrivals in each slot are independent of each other. Then $P(i)$, the probability that the first packet arrival occurs in the ith slot, is

$$P(i) = \underbrace{(1 - p)(1 - p)(1 - p)}_{i-1 \text{ times}} \ldots p \tag{2.35}$$

$$\boxed{P(i) = (1 - p)^{i-1}p \quad i = 1, 2 \ldots} \tag{2.36}$$

This is the geometric distribution.

Let's find the mean number of slots until the first arrival. Using first principles,

$$\bar{n} = \sum_{i=1}^{\infty} iP(i) \tag{2.37}$$

$$\bar{n} = \sum_{i=1}^{\infty} ip(1 - p)^{i-1} \tag{2.38}$$

As p is not a function at the summation index, i, one can bring p outside of the summation

$$\bar{n} = p \sum_{i=1}^{\infty} i(1 - p)^{i-1} \tag{2.39}$$

Multiply and dividing by $1 - p$ yields

$$\bar{n} = \frac{p}{1 - p} \sum_{i=1}^{\infty} i(1 - p)^{i} \tag{2.40}$$

$$\bar{n} = \frac{p}{1 - p} \sum_{i=0}^{\infty} i(1 - p)^{i} \tag{2.41}$$

Here $i = 1$ becomes $i = 0$ as the value of the zeroth term is zero. Using a summation from the appendix,

$$\bar{n} = \frac{p}{1 - p} \frac{1 - p}{p^2} \tag{2.42}$$

$$\boxed{\bar{n} = 1/p} \tag{2.43}$$

Thus, one can tabulate the average number of slots until the first arrival

p	$\mu = 1/p$
0.1	10
0.2	5
0.5	2
0.9	1.1
1.0	1.0

Intuitively these values make sense. For instance, if an arrival occurs with probability $p = 0.2$, it takes five slots on average until the first arrival.

To find the variance in the number of arrivals, from first principles,

$$\sigma^2 = \sum_{i=1}^{\infty}(i - \mu)^2 P(i) \tag{2.44}$$

where $P(i)$ is the geometric distribution.

One can show after a few steps

$$\boxed{\sigma^2 = (1 - p)/p^2} \tag{2.45}$$

Example: Let's consider a web server example. A major web site has a load balancing computer that feeds requests for web pages to 1 of 12 computers that store the web pages. "Load balancing" is done so that no single computer is overloaded. Let q be the probability that a computer can accept a web page request (the computer is not overloaded).

When it gets a web page request, the load balancing computer checks with each of 12 computers in turn (sequentially) to see whether it is overloaded. Find an expression for the probability that the ith computer it checks can accept the job (is not overloaded).

Solution: The situation is clearly modeled by a geometric distribution. The probability $P(i)$ that the ith computer accepts the request is

$$P(i) = (1 - q)^{i-1}q$$

Here $i - 1$ computers reject the request [with probability $(1 - q)^{i-1}$ and the ith computer accepts it (with probability q)]. Note that there is a finite probability that none of the computers accepts the request with probability

$$P(\text{reject}) = 1 - \sum_{i=1}^{12}(1 - q)^{i-1}q$$

or

$$P(\text{reject}) = \sum_{i=13}^{\infty} (1-q)^{i-1} q$$

Question 2: What is the the probability of n arrivals in N slots?

This question is answered by using a binomial distribution. Let's say that in N slots there are n arrivals. The arrivals are assumed to be independent of one another. The probability of n arrivals in N slots in a specific pattern of placement is

$$p^n (1-p)^{N-n} \tag{2.46}$$

The probability of n arrivals in *any* pattern of placement in N slots is the probability of a single pattern occurring times the number of patterns. Thus we need to multiply the probability expression above by the number of patterns involving n arrivals distributed among N slots (of course, $n \le N$). The number of patterns is

$$\binom{N}{n} \tag{2.47}$$

One then has the binomial distribution for the probability of n arrivals in N slots

$$\boxed{\binom{N}{n} p^n (1-p)^{N-n}} \tag{2.48}$$

A bit of thought will show that, for a given p, the average number of arrivals in N slots is Np.

Example A: Consider the previous web server example. Find an expression for the probability that six or more of the computers are overloaded. You may use a summation.

Solution to Example A: The probability $P_{\ge 6/12}$ that 6 or more of the 12 computers are overloaded is a sum of binomial probabilities

$$P_{\ge 6/12} = P(6 \text{ overloaded}) + P(7 \text{ overloaded}) + \cdots + P(12 \text{ overloaded})$$

$$P_{\ge 6/12} = \sum_{i=6}^{12} \binom{12}{i} (1-q)^i q^{12-i}$$

Alternatively

$$P_{\ge 6/12} = 1 - P(0 \text{ overloaded}) - P(1 \text{ overloaded}) + \cdots + P(5 \text{ overloaded})$$

$$P_{\geq 6/12} = 1 - \sum_{i=0}^{5} \binom{12}{i} (1-q)^i q^{12-i}$$

Example B: A fault-tolerant computer system in a Jupiter space probe uses majority logic and three independent computers. "Votes" are taken, and if at least two computers agree to activate thrusters, the thrusters are activated. Let p be the probability that a single computer makes the wrong decision. Find the probability that a vote is incorrect.

Solution to Example B: The probability that a vote is incorrect is the probability that two computers are wrong plus the probability that three computers are wrong. Using the binomial distribution

$$P(\text{vote incorrect}) = P(2 \text{ computers wrong})$$

$$+ P(3 \text{ computers wrong})$$

$$= \binom{3}{2} p^2(1-p) + \binom{3}{3} p^3(1-p)^0$$

$$= 3p^2(1-p) + p^3$$

If $p = 0.01$, the probability of an incorrect vote is

$$P(\text{vote incorrect}) = 0.000297 + 10^{-6} = 0.000298$$

This is a 30-fold improvement in the reliability of the system. Naturally, a complete reliability analysis will consider such additional factors as the reliability of the voting logic and the degree to which the computers are independent (i.e., overheating in the space probe may cause simultaneous failures).

Question 3: What is the distribution of the time until the kth arrival where $k \geq 1$?

The distribution in question is the Pascal distribution (Goodman 04).

To find the Pascal distribution, with some thought, one can see that the probability of k arrivals, in N slots $P_N(k)$ is

$$P_N(k) = P[A]P[B] \tag{2.49}$$

Here $P[A]$ is the probability of $k-1$ arrivals in $N-1$ slots. Also $P[B]$ is the probability of an arrival in the Nth slot.

Thus, using binomial style statistics (see above)

$$P[A] = \binom{N-1}{k-1} p^{k-1}(1-p)^{N-1-(k-1)} \tag{2.50}$$

$$P[B] = p \tag{2.51}$$

As $P[AB] = P[A]P[B]$, then

$$P_N(k) = \binom{N-1}{k-1} p^k (1-p)^{N-k} \quad k = 1, 2, 3 \ldots \qquad (2.52)$$

Naturally the mean number of slots holding k arrivals is $\mu = k/p$ [see equation (2.43)]. Again, the expression makes intuitive sense.

Example: A buffer is fed by a packet stream that can be modeled as a Bernoulli process. The buffer dumps its contents onto a network when the tenth packet arrives. What is the probability that the tenth packet arrives in the 30th slot?

Solution: Using the Pascal distribution with these parameters

$$P_{30}(10) = \binom{29}{9} p^{10}(1-p)^{20}$$

The mean number of slots holding ten packets is $10/p$.

2.4 Multiple Access Performance

2.4.1 Introduction

An early problem in computer networking was the multiple access problem. That is, how does one share a common medium (i.e., channel) among a number of users in a decentralized manner. One important early (circa 1980) application was the first version of Ethernet, the popular local area network protocol. Here the channel was a coaxial cable. A second earlier application was the Aloha packet radio network interconnecting the University of Hawaii. In the following we analyze the performance of two discrete time tractable Ethernet models and a continuous time Aloha model. In both cases, we will find that there is an intermediate value of the traffic load "offered" to the network (called offered load) that maximizes throughput.

2.4.2 Discrete Time Ethernet Model

The first implementation of Ethernet involved stringing a coaxial cable office to office in a linear fashion. Computers in each office tap into the cable, which serves as a "private" radio channel in the sense that transmissions are confined (electromagnetically) to the cable (see Figure 2.5).

The basic Ethernet protocol is called CSMA/CD, which stands for carrier sense multiple access with collision detection. It functions as follows. Any station sensing an idle channel will attempt to transmit as soon as it has a message (or packet). Although all stations will be able read the header of the

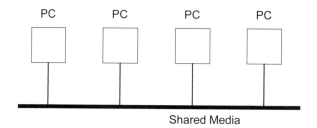

Fig. 2.5. Shared media Ethernet

packets, only the station whose address matches the destination address will pay attention to it.

In this sense, the original Ethernet cable served as a broadcast medium. The basic shortcoming of the CSMA/CD protocol is that if two or more stations sense the cable to be idle at about the same time, the two or more stations may launch packets onto the cable at the same time, the transmissions will overlap, and this will be unintelligible to the stations. This overlapping of transmissions is called a "collision."

Stations in the early 1980s implementations of Ethernet can detect a collision situation. The multiple stations involved in a collision will each reschedule their transmissions into the future hoping to transmit them without a collision.

It is possible to create an approximate discrete time model of Ethernet transmission using the binomial distribution. That we will do in this subsection. A more accurate model is discussed in the next subsection.

Let time be divided into equi-spaced slots. Each slot can hold a fixed length packet. Each of N stations will attempt to transmit a single packet in a specific slot with independent probability p (and not transmit with independent probability $1 - p$). Thus each slot holds 0 to N packets. However, useful information is conveyed only if there is a single transmission (packet). A "collision" occurs if two or more stations attempt to transmit in the same time slot. The channel is idle if no station transmits in a specific slot. Thus there are three possible mutually exclusive events (idleness, useful transmission, collision) that can occur in a specific slot. Using the binomial distribution, one has

$$P[0 \ xmssns] = (1 - p)^N \tag{2.53}$$

$$P[1 \ xmssn] = Np(1 - p)^{N-1} \tag{2.54}$$

To calculate the probability of two or more transmissions in a slot, one could sum the probabilities of each of n (i.e., $2 \le n \le N$) stations transmitting. A more clever approach is to realize that

$$P[\text{collision}] = 1 - P[0 \ xmssns] - P[1 \ xmssn] \qquad (2.55)$$

$$P[\text{collision}] = 1 - (1 - p)^N - Np(1 - p)^{N-1} \qquad (2.56)$$

Naturally we can plot each of these three probabilities versus p, the "offered load" to the network. Most interesting is the plot of useful throughput ($P[1 - xmssn]$) versus p. As p increases the throughput is initially linear in p (light load) and then saturates and decreases (heavy load). Thus a heavily loaded 10 Mbps Ethernet may only carry, say, 4 or 5 Mbps of actual traffic. The decrease in throughput is due to the increasing fraction of time wasted in collisions as p is increased. This the price paid for a completely decentralized system with no centralized scheduling.

We can find the value of p that maximizes throughput. Using our calculus knowledge, one can set the derivative of throughput ($P[1\,xmssn]$) with respect to p equal to zero, solve, and obtain

$$p_{\text{optimal}} = 1/N \qquad (2.57)$$

This makes intuitive sense. If N stations attempt to access the cable, each with probability $1/N$, the offered load is simply $N \times (1/N) = 1.0$, or one packet per slot on average. That is, beyond this match between offered load and normalized network capacity, more and more collisions result.

It was mentioned above that the binomial model is approximate. This is because of several implicit assumptions made for the binomial model that do not perfectly represent actual Ethernet operation. For instance, in reality, packets/frames are of variable length and can be transmitted at any (continuous) time instant. Second, the propagation delay of the coaxial cable should be modeled to adequately represent the collision process. Also in reality the probability a specific attempt to transmit in a slot is correlated with transmission failures in the recent past. Finally a station usually has buffers to hold packet waiting to be transmitted. However our binomial-based model does portray the key feature of CSMA/CD performance, the drop in throughput as load is increased.

The CSMA/CD protocol used in Ethernet is said to be 1 persistent. That is, if the channel is sensed as idle by a station with a packet to send, the station transmits instantly with probability 100% (i.e., 1 persistent). In a p persistent protocol the probability that a station with a packet transmits on an idle channel is $p \times 100\%$. With such less greedy stations, throughput actually does improve, although with a corresponding increase in delay. That is the delay experienced by each packet increases with the less greedy access policy.

A word is in order on the differences between the Ethernet cable (bus) and the typical "bus" inside a computer. Computer buses typically have a high throughput achieved by using a number of wires (say, 32 or 64 wires) in parallel. That is, for a 32 bit "wide" bus, each wire may operate at 100 Mbps so that the total capacity is 32×100 Mbps or 3.2 Gbps of capacity.

Local area network (e.g., Ethernet) "buses" on the other hand are essentially a single wire or a small number of wires. Given the large distances (10 to 100s of meters) involved, this makes economic sense.

Because computer buses have a small physical size, their use can be governed by a bus scheduler. A bus scheduler determines, in some fair manner, how much bus time to grant to each system (CPU, I/O..) requesting service. There are no collisions. An early Ethernet cable, on the other hand, has significant propagation delay due to its multimeter size. Thus at the time it was reasonable to trade-off wasted capacity due to collisions for the simplicity of no central scheduler.

In fact Ethernet performance degrades as physical network size is increased. This has led to some interesting trade-offs as Ethernet speed pushes above 10 Mbps, which are discussed below. The trend in the more recent versions of both Ethernet and other local area networks is to make use of a hub/star architecture.

In a hub architecture, stations are wired directly to a hub (switch), which can be the size of a cigar box or smaller, in a star type pattern. A number of stations wired to a hub may make use of a shared media card, which mimics the larger coaxial Ethernet connections, including the presence of collisions. In a switched Ethernet card, on the the other hand, buffers hold transmissions until they can be scheduled on a mini-computer bus in the hub. In this case there are no collisions. It should be noted that the trend toward hub architecture includes such local area network protocols as fast Ethernet (100 Mbps) and Gigabit Ethernet (1000 Mbps).

2.4.3 Ethernet Design Equation

With some more elaboration we can develop the basic Ethernet equation that predicts channel efficiency (i.e., utilization) as a function of data rate, minimum frame length, and network size.

Assume a heavy load of k stations attempting to access the shared media (cable). Following Tanenbaum 03, let p be the independent probability that a single station attempts to access the media. Then the probability only a single station acquires the channel A is given by a binomial distribution

$$A = kp(1 - p)^{k-1} \tag{2.58}$$

Now from the previous section we know A is maximized when $p = 1/k$. The probability that a "contention interval" has j slots is given by a geometric distribution

$$P(j) = P[j \text{ slot contention interval}] = A(1 - A)^{j-1} \tag{2.59}$$

Here the contention interval is the interval during which stations contend for access to the channel. Then the mean (average) number of slots per contention interval is

$$\sum_{j=0}^{\infty} jP(j) = \sum_{j=0}^{\infty} j(1 - A)^{j-1} A \tag{2.60}$$

$$= A \sum_{j=0}^{\infty} j(1 - A)^{j-1} \tag{2.61}$$

$$= \frac{A}{1 - A} \sum_{j=0}^{\infty} j(1 - A)^{j} \tag{2.62}$$

$$= \frac{A}{1 - A} \times \frac{1 - A}{A^2} \tag{2.63}$$

$$= \frac{1}{A} \tag{2.64}$$

Here we have used the same procedure used in simplifying some of the previous summations in this chapter. Again the slotted model we use is an approximation to what is really a continuous time system.

Let the propagation delay from one end of the cable to the other end be τ. The worst-case contention interval duration occurs when a station at one end of the cable transmits onto an idle channel, only to have the station at the opposite end of the cable transmit just before the first station's signal reaches it. In this case the time between the first station beginning transmission and finding out a collision occurs is 2τ (the round-trip propagation delay). In the worst case, then, each contention interval slot is 2τ seconds long.

The mean contention interval (consisting of $1/A$ contention interval slots) is thus $2\tau/A$ seconds. However with an optimal (throughput maximizing) choice of p, $1/A = e$ as $k \to \infty$. So $2\tau/A = 5.4\tau$.

Next an expression for channel efficiency (utilization) is needed. This is the ratio of the time useful information is transmitted to the total time it takes to transmit the information. If P is the frame (packet) length in seconds,

$$U = \frac{P}{P + \frac{2\tau}{A}} \tag{2.65}$$

But

$$P = \frac{F}{B} \tag{2.66}$$

Here F is the minimum frame size (in bits) and B is the data rate (in bits per second). Then

$$U = \frac{\frac{F}{B}}{\frac{F}{B} + \frac{2\tau}{A}} \tag{2.67}$$

But from just before, $2\tau/A = 2\tau e$, so

$$U = \frac{\frac{F}{B}}{\frac{F}{B} + 2\tau e} \tag{2.68}$$

$$U = \frac{1}{1 + \frac{2B\tau e}{F}} \tag{2.69}$$

However the one-way propagation delay is L/c, where L is the cable length (in meters) and c is the speed of light or electromagnetic radiation (in meters per second). Thus

$$U = 1/\left(1 + \frac{2BLe}{cF}\right) \tag{2.70}$$

This is the basic Ethernet design equation. The original Ethernet standard (IEEE 802.3) produced reasonable utilizations with a 10 Mbps data rate, 512 bit minimum frame size, and maximum cable size of 500 to 1000 meters.

The challenge to the Ethernet community over the years was how to boost data rate while maintaining utilization. Clearly if B is simply increased, U will drop. Thus in producing Fast Ethernet with a 100 Mbps data rate in the early to mid-1990s, B was increased by a factor of ten and the maximum network size L was reduced by a factor of ten (to about 50 meters), so the product BL in the equation is constant.

In designing gigabit (1000 Mbps) Ethernet during the late 1990s, this trick couldn't be repeated since network size L would be an unrealistic 5 meters. Instead B was increased by a factor of ten and F, the minimum frame size, was increased by a factor of eight (from 512 bits to 512 bytes). Thus the ratio B/F is approximately constant and utilization levels are largely maintained.

Finally, note that utilization increases as frame size increases. This is because for longer frames, collisions are a smaller portion of the transmission time of a packet.

2.4.4 Aloha Multiple Access Throughput Analysis

In the second look at sharing a communication medium, the Aloha packet radio network developed circa 1970 for the University of Hawaii is examined. The idea at the time was to connect the Hawaiian island campuses with a distributed radio network. The basic layout appears in Figure 2.6. The central station monitors a single incoming channel of packets from the satellite islands. What the central station "hears" on the incoming channel is broadcast back to all satellite islands on a second channel.

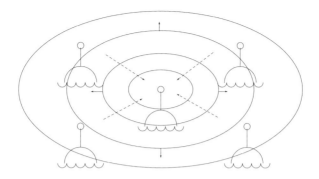

Fig. 2.6. Aloha network geography (topology)

There is no central control in this system. A station on a satellite island simply broadcasts into the main island in the hopes it is the only incoming transmission at that time. That is, if two or more stations transmit toward the central repeater at about the same time, the messages will overlap and be unintelligible. This is the Aloha equivalent of an Ethernet collision. Individual satellite stations monitor the outbound channel to hear what the central repeater hears (an intelligible transmission, unintelligible transmission, or idleness).

However, without collision detection and with the relatively larger transmission distances, the performance of Aloha is lower than that of Ethernet for the same channel speed. Intuitively, though, one would expect the type of performance of Aloha to be similar to Ethernet. That is, as offered load is increased, a linear growth that is throughput first saturates and then decreases due to collision-like behavior.

The standard analysis (Abramson 70, 85, Roberts 72, Saadawi 94) to be used to measure this performance is a bit different in the details compared with the Ethernet-like analysis of the previous section.

Now consider Figure 2.7. In the channel system of Figure 2.7 there are two streams of traffic attempting to enter the channel, S packets per unit time T of new, first-time traffic and G packets per unit time T of new traffic and repeated transmissions that didn't get through on earlier attempts. Let the

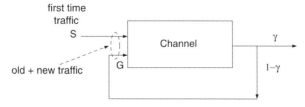

Fig. 2.7. Aloha channel diagram

probability of a successful transmission be γ. Since the throughput in a stable channel equals the outflow (or inflow),

$$S = G\gamma \tag{2.71}$$

But what is γ? There are two related scenarios here. In the first scenario, known as slotted Aloha, time is broken into equi-spaced slots of duration T seconds per slot. One packet fits in one slot. Collision-like behavior occurs when more than one packet attempts to be transmitted in the same slot. We assume that the probability $P(k)$, that there are k packets transmitted in a specific slot, is Poisson. That is, the arrivals in one slot time follow a Poisson distribution. This is reasonable as the transmissions from the satellite islands should be independent of each other, at least to a first approximation.

$$P(k) = \frac{(Gt/T)^k}{k!} e^{-(G/T)t} \tag{2.72}$$

Here G/T is the equivalent of the packet arrival rate, with overall units of the number of packets in an interval t. Here also we use a "tagged" packet approach to estimate throughput. That is we observe one given (tagged) packet and observe its chance of being successfully transmitted. For our tagged packet to be successful, there must be no other new or old transmissions ($k = 0$). Thus

$$\gamma = e^{-(G/T)t} \tag{2.73}$$

Here small t is the window of vulnerability in seconds. That is, it is the time period when a packet(s) other than the tagged packet may attempt to transmit, causing a collision. For slotted Aloha, clearly $t = T$ (the slot width and packets transmission time). Thus

$$\gamma = e^{-(G/T)T} \tag{2.74}$$

and

$$\gamma = e^{-G} \tag{2.75}$$

$$\boxed{S = Ge^{-G}} \tag{2.76}$$

Plotting throughput (S) versus offered traffic (G), one finds (Figure 2.8) throughput is maximized at a value of 36.8% at $G = 1.0$.

The second possible scenario is called "pure" Aloha. Here packets may arrive at any time instant If one considered partially overlapping packets leading to collision, then the window at vulnerability during which a tagged packet

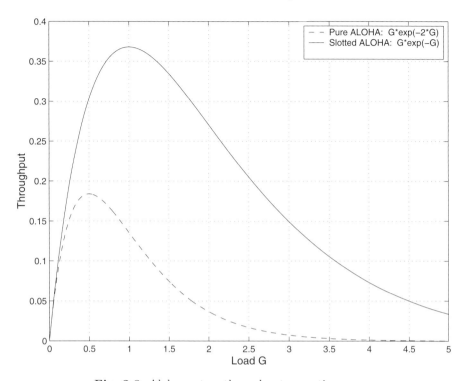

Fig. 2.8. Aloha system throughput operating curves

is susceptible to an overlapped transmission is, with some thought, $t = 2T$. Then

$$S = Ge^{-2G} \tag{2.77}$$

Now throughput is maximized at a value of 18.4% at $G = 0.5$. Although this is significantly lower than the 36.8% of slotted Aloha, there is no need for slot timing boundaries. A key implicit assumption in our look at Aloha is that G, the arrival stream of new and old traffic, is Poisson. It is actually clearer that this would be true for the new distributed traffic stream S than for G. This is because, over relatively short periods of time, new incoming transmissions are independent of each other. Actually G is correlated with S. That is, a failed transmission attempt increases the rate of future attempts. This correlation aside, the performance evaluation we have, is reasonable for a first look at Aloha. See Rom 90 for a more detailed treatment.

2.4.5 Aloha Multiple Access Delay Analysis

With some work we can determine the average delay experienced by packets attempting to transit an Aloha channel. Let's look first at slotted Aloha. If a packet is not successful on an attempt through the channel, it is randomly rescheduled some time into the future. Of course it may take a number of attempts to get a packet through the channel, particularly if the load is heavy.

Let's say a packet that needs to be rescheduled is uniformly likely to be transmitted over K slots (in each with probability $1/K$). Then following Saadawi 94, the average number of slots a packet waits before transmitting is $\bar{i}$

$$\bar{i} = \sum_{i=0}^{K-1} i \frac{1}{K} \tag{2.78}$$

$$\bar{i} = \frac{1}{K} \sum_{i=0}^{K-1} i \tag{2.79}$$

$$\bar{i} = \frac{1}{K} \frac{K(K-1)}{2} \tag{2.80}$$

$$\bar{i} = \frac{K-1}{2} \tag{2.81}$$

Here we use $K - 1$ in the summation rather than K as we are interested in the time a packet waits before transmitting (0 to $K - 1$ slots at most).

If T is the duration of a slot in this "backoff" algorithm, then the average backoff time is $\bar{i}T$.

Next, the time cycle duration of an unsuccessful attempt consists of three components

$$T_u = 1 + R + \frac{K-1}{2} \tag{2.82}$$

One can see this includes the packet transmission time (1), propagation delay (R), and the average backoff time $[(K - 1)/2]$. With algebra

$$T_u = R + \frac{K+1}{2} \tag{2.83}$$

Here the units are in slots. Multiplying by the number of seconds/slot T, one has

$$T_u = T \left[R + \frac{K+1}{2} \right] \tag{2.84}$$

Here T_u, the average unsuccessful cycle time, is in seconds.

Now let E be the average number of retransmissions. Then the average delay (in slots) to get a packet through the channel D has three components

$$D = \frac{1}{2} + \frac{T_u}{T}E + (1 + R) \tag{2.85}$$

Here $1/2$ is the average slot time difference between a packet becoming ready to send and actually being sent (uniformly on slots of normalized duration 1). The second term is the time for unsuccessful transmissions (measured in slots). The last term is the time to successfully transmit the packet plus propagation delay (R). Here also propagation delay is the time for the radio signal to physically transmit the channel.

Naturally we need to calculate E. From the previous section we know the probability a packet successfully transmits the Aloha channel is e^{-G}. Using the geometric distribution, the probability a packet transmits the channel in n attempts $P(n)$ is

$$P(n) = (1 - e^{-G})^{n-1}e^{-G} \tag{2.86}$$

The average number of attempts, including the successful one, is then (see section 2.3)

$$\bar{n} = \sum_{n=1}^{\infty} nP(n) \tag{2.87}$$

$$\bar{n} = \sum_{n=0}^{\infty} n(1 - e^{-G})^{n-1}e^{-G} \tag{2.88}$$

$$\bar{n} = e^{G} \tag{2.89}$$

Then substituting

$$E = e^{G} - 1 \tag{2.90}$$

Here we subtract the successful last attempt from the total number of attempts to find E, the average number of unsuccessful attempts E. We can also say (as $S = Ge^{-G}$ from the previous subsection)

$$E = \frac{G}{S} - 1 \tag{2.91}$$

If one plots D (average delay) versus G (total throughput) for slotted Aloha, one obtains an exponential curve as in Figure 2.9. For low or medium load, average delay is small. However as one approaches a fully loaded channel, delay increases exponentially. This type of behavior is similar in spirit to multiple-access resource-sharing systems such as web servers and Markovian queues (see the next chapter).

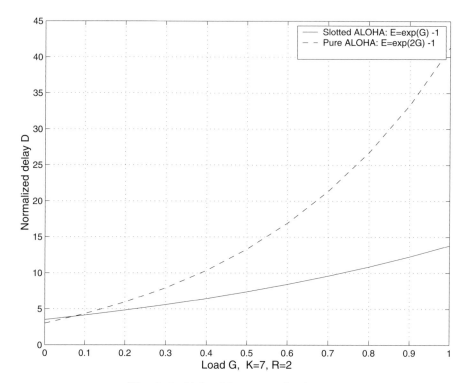

Fig. 2.9. Aloha delay versus load curves

In any case, putting everything together, one has

$$D = \tfrac{1}{2} + \left[R + \tfrac{K+1}{2}\right]\left[e^{G} - 1\right] + [1 + R] \qquad (2.92)$$

For pure Aloha, the delay equation, using a similar approach as that for slotted Aloha, is; in slots

$$D = \frac{T_u}{T}E + (1 + R) \qquad (2.93)$$

Since a packet is transmitted as soon as it is ready in pure Aloha, there is no average submission time delay "1/2" term. Note here $E = e^{2G} - 1$. Thus

$$D = \left[R + \tfrac{K+1}{2}\right]\left[e^{2G} - 1\right] + [1 + R] \qquad (2.94)$$

For the parameters chosen in Figure 2.9, it can be seen that pure Aloha has significantly more delay at heavy loads than slotted Aloha. This is a consequence of pure Aloha's e^{2G} term, versus slotted Aloha's e^{G} term.

2.5 Teletraffic Modeling for Specific Topologies

2.5.1 Introduction

Some teletraffic applications involve the use of specific topologies. A network topology is a particular geometric arrangement of links and nodes. Certain specific topologies have a structure that aids in teletraffic modeling.

In the following subsection, teletraffic modeling for linear networks, trees, and networks over circular regions is discussed. Uniform traffic will be assumed in all of these topologies. That is, the amount of traffic between any pair of nodes is the same, either on average or in fact. Thus the loadings we observe at particular points in each topology are, respectively, either average or deterministic quantities.

These example topologies are meant to be representative.

2.5.2 Linear Network

Let's investigate the linear topology of the DQDB metropolitan area network. A metropolitan area network is meant to be a high-speed backbone for smaller local area networks. The DQDB protocol (Distributed Queue Dual Bus), developed originally in Australia in the 1980s, was chosen as the IEEE 802.6 metropolitan area network (Garrett 90, Rodrigues 90).

The actual network logical topology consists of a dual bus architecture. That is, assume the stations $1, 2, \ldots N$ are placed in a straight line (linear array). Each station connects to two fiber-optic buses. The upper bus handles only transmissions from left to right (from station i to $i + 1$, $i + 2 \ldots$). The lower bus handles transmissions from right to left (from station i to $i - 1$, $i - 2 \ldots$). Thus there is only one route between each pair of stations (i.e., Figure 2.10).

Let's assume a uniform traffic scenario. That is, each node is equally likely to direct a packet to any other station with the same constant packet generation rate. With this homogeneous traffic arrival model and the network

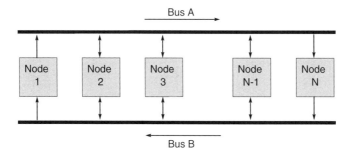

Fig. 2.10. DQDB network topology

symmetry, it can be seen that half of the average network traffic uses the
upper bus and the other half uses the lower bus. Thus we can consider the
upper bus in isolation.

The upper bus carries traffic from left to right (station i to $i+1, i+2\ldots$).
On both buses time is divided into equal width slots. One slot holds one
packet. In order to provide approximate first-in first-out (FIFO) service to
packets in the DQDB network, each station maintains a counter for each bus.

When a station wishes to transmit on the upper bus (left to right), it sets
the next available request bit in a packet on the lower bus (which moves to
the stations to the left of station i). The counter at station i maintains the
difference between the number of idle (empty) slots passing the station on the
upper bus and the outstanding number of packets that need to use slots on
the upper bus originating from stations $i + 1, i + 2, \ldots$. The second element
of the difference is obtained from the number of (set) request bits passing the
station on the lower bus. Once the difference is zero, station i may launch a
packet on the upper bus. Thus the counter allows a station to delay launching
a packet on the upper bus until earlier demand for slots on the upper bus
from stations to the right of the station has been satisfied.

Thinking about this protocol, one can see that it, in its basic implementation, provides approximate first-in first-out service over the DQDB network
without any central scheduling mechanism. It is only approximately FIFO
as there is only one request bit per packet, transmission speed is finite, and
propagation delay is nonzero. All of this slows the diffusion of information on
outstanding packets and idle slots in the network.

A limiting factor to the traffic capacity of a DQDB network is because
in a basic implementation each slot carries a packet between a single pair of
stations. That is, if a packet is sent from i to j on the upper bus, ideally station
j should erase the packet once it is read, allowing stations downstream on the
upper bus to "reuse" the slot. However, "active" erasure stations capable of
erasing a slot's packet introduce unwanted delay in reading the packet header
and optical electrical/optical conversion.

A proposed compromise is to use only a small number of active erasure
stations in the network. The design question is then where to place active
erasure stations and how many of them to deploy. Garrett 90 provide a solution
for both problems. We will only consider the question here of what is the
theoretical maximum boost if every station is an active one.

Here we will make explicit use of the linear network topology. Assume that
the upper bus can be modeled as a line segment from 0 (left most station) to
1 (right most station). The amount of traffic (transmission density) generated
at a location without erasure stations (on $[0, 1]$) is given by

$$f_T(x) = 2\gamma(1 - x) \quad 0 \le x \le 1 \tag{2.95}$$

This relationship is linear. The closer a node is to the right terminus, the less traffic it generates on average since there are fewer stations to the right of it.

Here γ is the single bus maximum throughput. The probability that a slot at location x is occupied is, with some calculus

$$F_O(x) = \int_0^x f_T(\chi)d\chi = \gamma x(2 - x) \tag{2.96}$$

A fully utilized DQDB network has 100% occupancy at the right terminus, which implies that with $F_O(1) = 1$ that $\gamma = 1$ for the network without slot reuse.

Now, consider how much traffic is received at location x (reception density)

$$f_R(x) = 2\gamma x \quad 0 \le x \le 1 \tag{2.97}$$

That is, the closer to the right terminus a station is, the more traffic it receives (as there are more stations to the left of it sending traffic on the upper bus).

Finally the occupancy distribution at location x on the network is, assuming "destination release (DR)" or erasure at every station, is

$$F_{O,\text{DR}}(x) = \int_0^x (f_T(\chi) - f_R(\chi))d\chi \tag{2.98}$$

$$\boxed{F_{O,\text{DR}}(x) = 2\gamma x(1 - x)} \tag{2.99}$$

On the interval $[0, 1]$ this occupancy distribution is maximized at $x = 0.5$, in the network center. That is, setting $F_{O,\text{DR}}(\gamma) = 1$ yields $\gamma = 2$. In other words, the maximum theoretical DQDB throughput boost with every station implementing erasures is 100% (a factor of 2).

Garrett 90 go on to determine a 50% boost using a single erasure station at the network middle ($x = 0.5$) and the optimal locations and throughput boosts for $n = 2, 3 \ldots$ erasure stations. In fact the good news is that the use of a relatively small number of erasure stations yields a performance close to the theoretical maximum. See Garett 90, Rodrigues 90, or Robertazzi 00 for more detailed treatments.

2.5.3 Tree Networks

Tree topologies can be modeled as a graph with no loops (cycles). Usually one node is identified as a root. Trees are important as they can be embedded into arbitrary interconnection networks (graphs) to provide connectivity between nodes. For a set of N nodes, a spanning tree connects all nodes.

Practically, cable TV systems and hierarchical circuit-switching telephone networks are tree networks.

Tree networks are also related to multiplexing hierarchies. Consider a digital telephone network. A digital phone can produce 64 kbps of uncompressed voice (i.e., 8 bits/sample × 8000 samples/second). In the North American phone system, 24 voice calls are interleaved to produce a 1.544 Mbps "T1" stream. Four T1 streams can be interleaved into a 6.312 Mbps T2 stream and on and on. Demultiplexing proceeds in the opposite manner. That is, a T2 stream is broken down into constituent T1 streams and each T1 stream is broken into 24 phone channels.

In the following discussion, two tree type problems are considered. The first problem involves capacity calculation. The second problem deals with average path distance.

Capacity Calculation

Consider a binary tree network as in Figure 2.11. A binary tree has two children nodes per node. In this capacity allocation problem, assume that the actual users are in the bottom level. The other nodes are switching elements (i.e., multiplexers and demultiplexers M/D).

Assume all traffic passes through the root. Suppose that the total capacity between each pair of bottom-level nodes, including both directions, is one unit of capacity. How much capacity is needed in link X (the link just below the root on the left side)?

Intuitively one can see that the closer a link is to the root, the more capacity is needed. More specifically traffic passing through link X consists of two components. One component is traffic from the lower left user nodes to the lower right nodes. If there are N lower level nodes, this traffic, with the usual uniform loading assumption, is $(N/2 \times N/2)$ or $(N/2)^2$. That is, the first component is the number of bottom-level nodes on the left side of the tree

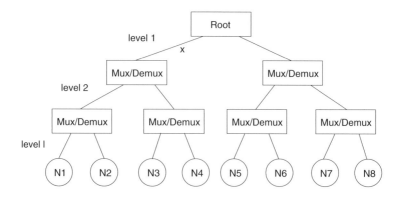

Fig. 2.11. Tree network

times the number of bottom-level nodes on the right side of the tree. This is because each pair of bottom-level nodes is equally likely to generate one unit of traffic. The second component of traffic at link X is traffic from one lower left node to one of the other lower left nodes $(N/2) \times ((N/2) - 1)$. The term of "-1" occurs because a node does not send traffic to itself.

Thus

$$\text{Load}_x = \left(\frac{N}{2}\right)^2 + \left(\frac{N}{2}\right)\left(\frac{N}{2} - 1\right) \tag{2.100}$$

Note that for an l-level tree, there are 2^l nodes in the lower level or

$$N = 2^l \tag{2.101}$$

Substituting this equation into the previous one yields

$$\text{Load}_x = \left(\frac{2^l}{2}\right)^2 + \left(\frac{2^l}{2}\right)\left(\frac{2^l}{2} - 1\right) \tag{2.102}$$

$$\text{Load}_x = (2^{l-1})^2 + (2^{l-1})(2^{l-1} - 1) \tag{2.103}$$

$$\text{Load}_x = 2^{2l-2} + (2^{2l-2} - 2^{l-1}) \tag{2.104}$$

$$\text{Load}_x = 2^{2l-1} - 2^{l-1} \tag{2.105}$$

One can see here that the traffic load at link X grows as the square of N or approximately two to an exponent of $2l - 1$.

Assume now that shortest path routing is used in the tree of Figure 2.11. That is, traffic is not necessarily routed through the root but through the shortest path involving M/D.

In this case the traffic load in link X consists only of the traffic from the lower left nodes to the lower right nodes (that must transit the root). Thus

$$\boxed{\text{Load}_x = (N/2)^2 = \left(2^l/2\right)^2 = (2^{l-1})^2 = 2^{2l-2}} \tag{2.106}$$

If there are T bps of traffic between each pair of nodes, the above load equation is valid if multiplied by T bps. Naturally, expressions for the load at any level may also be calculated in a similar manner. The fact that trees needed more capacity near the root under a uniform traffic assumption has led to the proposal for the use of "fat" trees as in Lieserson 85.

Distance Calculation

Consider a binary tree now where each node is a user, except for the root. Assume all transmissions must pass through the root. The question in the distance calculation problem to be considered is to find the average distance (number of hops) between any pair of nodes for a very large tree. A "hop" is an approximate measure of distance. In transiting a link between two neighboring nodes, one makes one hop.

To solve this problem (Tanenbaum 02) one can see that the average distance between nodes comprises two equal components. The first component is the average distance from the nodes to the root. The second component is the average distance from the root to the nodes. If we solve for the first component, we need only double it for the complete answer

$$\overline{\text{Distance}} = 2 \times \overline{1 \text{ way distance}} \tag{2.107}$$

A little thought will show that for a large binary tree about 50% of the nodes are in the lower most level, about 25% of the nodes in the next lower level, about 12.5% are in the next lower level, etc.

Thus

$$\overline{1 \text{ way distance}} = 0.5 \times l + 0.25 \times (l-1) + 0.125 \times (l-2)\ldots \tag{2.108}$$

Here we have a weighed sum of the fraction of nodes at each level and the levels distance to the root. This weighed sum can be written as

$$\overline{1 \text{ way distance}} = \sum_{i=0}^{l}(0.5)^{i+1}(l-i) \tag{2.109}$$

$$\overline{1 \text{ way distance}} = \sum_{i=0}^{l}(0.5)^{i+1}l - \sum_{i=0}^{l}i(0.5)^{i+1} \tag{2.110}$$

Each summation can be solved separately. We have (using a summation formula from the appendix) and letting $l \to \infty$

$$\sum_{i=0}^{l}(0.5)^{i+1}l = 0.5l\sum_{i=0}^{l}(0.5)^{i} \tag{2.111}$$

$$\sum_{i=0}^{l}(0.5)^{i+1}l = 0.5l\frac{1 - 0.5^{l+1}}{1 - 0.5} \tag{2.112}$$

$$\lim_{l\to\infty}\sum_{i=0}^{l}(0.5)^{i+1}l = l \tag{2.113}$$

For the second summation

$$\sum_{i=0}^{l} i(0.5)^{i+1} = 0.5 \sum_{i=0}^{l} i(0.5)^i \qquad (2.114)$$

Letting $l \to \infty$ and using another appendix summation formula

$$\lim_{l \to \infty} 0.5 \sum_{i=0}^{l} i(0.5)^i = 0.5 \frac{0.5}{(1 - 0.5)^2} = 1 \qquad (2.115)$$

Thus

$$\overline{1 \text{ way distance}} = l - 1 \qquad (2.116)$$

That the average distance from any node to the root is $l-1$ makes intuitive sense as 75% of the nodes are in the two lower most layers (i.e., 50% + 25%). Finally

$$\boxed{\overline{\text{Distance}} = 2 \times \overline{1 \text{ way distance}} = 2l - 2} \qquad (2.117)$$

2.5.4 Two-dimensional Circular Network

Let's consider a problem in the spatial distribution of stations and network intensity as a function of location. Specifically assume stations are uniformly distributed over a circular area of radius A. Let's also assure that two-way communication between nodes is homogeneous. That is, every pair of stations is likely to generate (an equal amount of) traffic. Finally assume shortest (geographic) paths are always used. That is, the communication path between two stations is a direct straight line.

The question then is what is the traffic intensity (amount) at any location in the circular area. An answer is provided by Kim 00. Intuitively, it can be seen that, because of symmetry, the intensity at a point in the circular area should be a function of its distance from the center.

It is assumed that the network covers a circular region with radius R. There are two equations for calculation of network traffic density that are the line $y = ax + b$ and the circular network boundary equation $x^2 + y^2 = R^2$. Traffic always follows a shortest path (straight line) route. To calculate the traffic "intensity" at an arbitrary point Z in the circular region, one can place a line through the point. Then there is a one-dimensional problem involving traffic generated between pairs of points on either side of Z on the line and passing through Z. The line is rotated 180 degrees about Z, and the intensity at Z is integrated. Figure 2.12 can be referred to for the following steps.

As mentioned for calculating the amount of traffic at an arbitrary point (x, y), the distances of $d1$ and $d2$ must be found. By multiplying these two distances, one obtains an amount of traffic intensity along the linear network

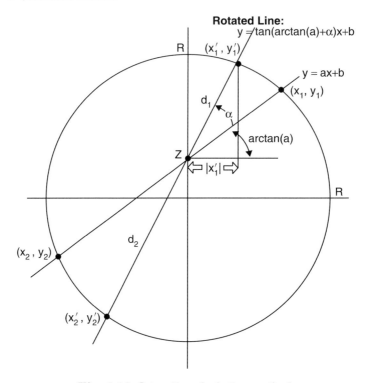

Fig. 2.12. Intensity calculation method

passing through Z. This amount varies depending on the line position. Without loss of generality, it is assumed that $Z = (0, b)$ on the line segment from $(0, 0)$ to $(0, R)$

$$d_1 = \frac{|x_1'|}{\cos(\alpha + \arctan(a))}, \quad d_2 = \frac{|x_2'|}{\cos(\alpha + \arctan(a))} \quad (2.118)$$

The rotated line that has a rotational angle α is found as (see Figure 2.12)

$$y = \tan(\arctan a + \alpha)x + b \quad (2.119)$$

In order to get the crossing points between the rotated line and the circle, substitute the above equation into $x^2 + y^2 = R^2$. Then

$$A = (\arctan a + \alpha)$$

$$x_{1,2}' = \frac{-b\tan(A) \pm \sqrt{b^2 \tan^2(A) - (1 + \tan^2(A))(b^2 - R^2)}}{1 + \tan^2(A)}$$

$$= \frac{-b\tan(A) \pm \sqrt{R^2 \tan^2(A) + R^2 - b^2}}{1 + \tan^2(A)} \quad (2.120)$$

Note that Z is assumed to be at $(0, b)$. Then the traffic intensity I_{linear} at $(0, b)$ along the linear component of the network is

$$I_{\text{linear}} = 2 \times d_1 \times d_2 \tag{2.121}$$

$$= 2 \frac{1}{\cos^2(A)} |x_1'||x_2'|$$

$$= 2 \frac{1}{\cos^2(A)} \left| \frac{b^2 \tan^2(A) - R^2 \tan^2(A) + b^2 - R^2}{[1 + \tan^2(A)]^2} \right|$$

$$= 2 \frac{1}{\cos^2(A)} \left| \frac{(b^2 - R^2) \tan^2(A) + b^2 - R^2}{[1 + \tan^2(A)]^2} \right|$$

$$= 2 \frac{1}{\cos^2(A)} \left| \frac{b^2 - R^2}{\frac{1}{\cos^2(A)}} \right|$$

$$= 2 |b^2 - R^2|$$

$$= 2(R^2 - b^2) \quad \text{since } R \geq b \tag{2.122}$$

This result implies that the traffic intensity along the line is independent of its rotational angle. The total traffic intensity I_{total} at $Z = (0, b)$ is the integration of the above equation by α, which is varying from 0 to π centered at $(0, b)$; that is

$$I_{\text{total}} = \int_0^\pi 2(R^2 - b^2) d\alpha \tag{2.123}$$

$$\boxed{I_{\text{total}} = 2\pi(R^2 - b^2).} \tag{2.124}$$

We need only integrate from 0 to π (not 2π) as the factor of 2 assumes two-way communication between each pair of points. The density of traffic inside the circular network is a quadratic function with a maximum of πR^2 at $b = 0$ (network center) and zero intensity at the boundary. It is somewhat surprising that the answer is so simple.

Figure 2.13 shows the quadratic traffic intensity as one moves from the boundary through the center of the circular network and out toward the opposite boundary. Also, the simulated result from Abernathy 91 is consistent with the traffic distribution found here.

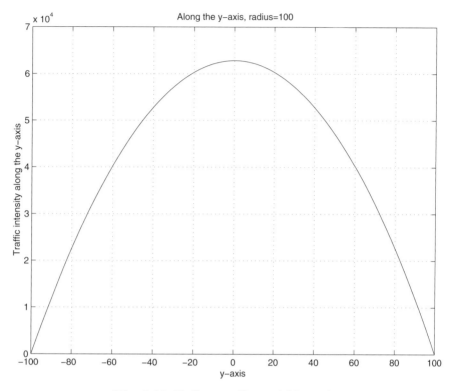

Fig. 2.13. Uniform traffic spatial intensity

2.6 Switching Elements and Fabrics

2.6.1 Introduction

How does one design a high-speed packet switch? The switching architecture with the best performance is space division switching. That is, one uses a multiplicity of relatively simple "switching elements" tied together by a structured interconnection network. The term "space division" comes from the fact that individual switching elements and paths are spatially separated. The term also allows a contrast to the older technology of time division switching (Saadawi 94).

An important motivation for the use of space division switching is the potential to implement this architecture in VLSI. The ability to simply copy (replicate) the same switching element many times on a chip speeds implementation. Also specialized VLSI chips have the potential to process many more packets per second than a general-purpose computer.

In the following sections we first consider fundamental and representative statistical models of switching elements. Crossbar interconnection networks are then discussed. Finally, a multiple bus system is studied.

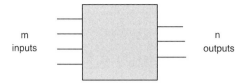

Fig. 2.14. $m \times n$ switching element

2.6.2 Switching Elements

Consider the $m \times n$ switching elements of Figure 2.14. There are m inputs and n outputs. Input and output processes are time slotted in nature. All slots have the same duration, and one packet fits into one slot exactly. Each input's packet arrival process is a Bernoulli process. The independent probability of a packet arriving in a given slot at a given input is p. The independent probability of no packet being in a slot at an input is $1 - p$. Arrivals are independent of each other from slot to slot and input to input.

Let's assume that time slot boundaries are synchronized. That is, the time slot boundaries for all inputs occur at the same time instants. The same is true for the outputs.

A switching element may implement one of a number of routing policies (for transferring incoming packets to output ports). Let's consider two representative policies.

Policy A: Say one has a 3×2 switching element. If one or two packets arrive in a slot on the inputs, the packets are sent to the output ports. It's not important here which of two arriving packets goes to which output port as long as only one of the two packets goes to each output port. If three packets arrive, one packet is randomly and fairly selected to be dropped (erased) and the other two packets proceed to the output ports.

Let's now look at some performance measures. A performance measure is a quantity that represents the performance of the system under consideration. Throughput, delay, blocking probability, and loss probability are common network performance measures. From the binomial distribution the probability of zero, one, two, or three arrivals across all inputs in a slot is

$$P(0 \text{ arrivals}) = \binom{3}{0} p^0 (1-p)^3 = (1-p)^3 \tag{2.125}$$

$$P(1 \text{ arrival}) = \binom{3}{1} p^1 (1-p)^2 = 3p(1-p)^2 \tag{2.126}$$

$$P(2 \text{ arrivals}) = \binom{3}{2} p^2 (1-p) = 3p^2(1-p) \tag{2.127}$$

$$P(3 \text{ arrivals}) = \binom{3}{3} p^3 = p^3 \tag{2.128}$$

The average (mean) throughput or flow per slot for the switching element can now be simply found. It is a a weighted combination of the number of arriving packets successfully going to the output ports and the probability of each number of packets arriving. Thus

$$\overline{\text{Throughput}} = 1 \cdot P(1 \text{ arrival}) + 2 \cdot P(2 \text{ arrivals}) + 2 \cdot P(3 \text{ arrivals}) \quad (2.129)$$

$$\overline{\text{Throughput}} = 1 \cdot 3p(1-p)^2 + 2 \cdot 3p^2(1-p) + 2p^3 \quad (2.130)$$

$$\overline{\text{Throughput}} = 3p(1-p)^2 + 6p^2(1-p) + 2p^3 \quad (2.131)$$

Here the weight multiplying $P(3 \text{ arrivals})$ is 2, and not 3, as if there are three arrivals only, two packets proceed to the two output ports.

Finally consider the probability that a packet is dropped. There are two ways to look at this situation. One is to take a "birds eye" view of the switching element. The probability that a packet is dropped is then simply the probability of three arrivals.

The second way to look at this situation is from the viewpoint of a "tagged" arriving packet. That is, suppose we take a seat on an incoming packet. If we know this packet is arriving on an input in the current slot with probability one, what is the probability a packet is dropped? It is equal to the probability that there are two additional arrivals on the other inputs. By way of contrast, the probability that the arriving tagged packet is dropped is one third of the probability that there are two additional arrivals as any of three arriving packets is equally likely to be dropped.

Policy B: Again consider a 3×2 switching element with three inputs and two outputs. Now suppose that an arriving packet is equally likely to go to either output port. Thus zero, one, two, or three packets may each prefer to go to a given output/port in a given slot. However, under the protocol to be examined, at most one packet may exit an output in one slot because of the limited capacity of the output link and the next switching element to which it leads. If more than one packet wants to go to a particular output port, one packet is chosen randomly and fairly to go to the output and the remaining packets are dropped.

Now the average throughput of a switching element is a weighted combination of the probability that a given number of arrivals occurs in a slot and the probability that a certain pattern of packets prefers to go to each output port in a slot for a given number of switching element arrivals, and the number of arriving packets successfully going to the output ports in a slot. The first term is the same as for policy A. Let us find the other terms.

Consider the probability that a certain number of packets were to go to a given output port in a slot for a given number of switching element arrivals. To find this probability consider Figure 2.15. Here there are two arriving packets to a 3 input 2 output (3×2) switching element. Call the two arriving packets A and B. The four boxes are for the patterns of packet output port preference. Each box is divided into two boxes, one for each of the two output ports.

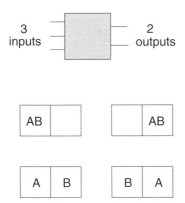

Fig. 2.15. Packet arrival patterns for two arriving packets to a 3 × 2 switching element

It can be seen that 50% of the time the packets go to the same output port and 25% of the time both packets prefer a specific output port. With two packets preferring the same output port, only a single packet actually leaves the output port under the assumed protocol so that the mean throughput has the term

$$1 \cdot \left(\frac{1}{2}\right) \cdot 3p^2(1-p) \tag{2.132}$$

The other 50% of the time when two packets arrive to the switching element each packet goes to a different output port (see Figure 2.15). Thus two packets will exit the output ports. So one has another term for the mean throughput of

$$2 \cdot \left(\frac{1}{2}\right) \cdot 3p^2(1-p) \tag{2.133}$$

Another way to find the factor of 1/2 intuitively is to realize that, if two packets arrive to the switching element, one will go to some output port. There is then a 0.5 probability that the second packet goes to the other output port, causing one packet to appear on each output port.

Now suppose that three packets arrive to the switching element. The possible patterns of the arriving packets' output port preferences are shown in Figure 2.16. There are eight possibilities. Call the arriving packets A, B, and C. One can see in the figure that there is 0.25 probability that all the arriving packets prefer the same output port. Naturally this leads to one departure. Thus the average throughput has the term

$$1 \cdot \left(\frac{1}{4}\right) \cdot p^3 \tag{2.134}$$

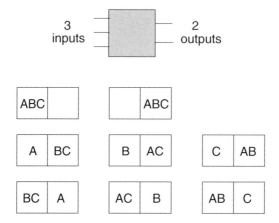

Fig. 2.16. Packet arrival patterns for three arriving packets to a 3×2 switching element

From the figure there is a 0.75 probability that two arriving packets prefer one output port and the third arriving packet prefers the other output port, leading to two packet departures from the switching element. Thus another average throughput term is

$$2 \cdot \left(\frac{3}{4}\right) \cdot p^3 \tag{2.135}$$

Finally, if only one packet arrives to the switching element, it definitely gets to the output port it prefers so that the final mean throughput term is

$$1 \cdot 3p(1 - p)^2 \tag{2.136}$$

Putting the five terms together, one has the following expressions for the mean throughput of policy B:

$$\overline{\text{Throughput}} = \underbrace{1 \cdot 3p(1 - p)^2}_{1-\text{arrival}} \tag{2.137}$$

$$= \underbrace{1 \cdot \left(\frac{1}{2}\right) \cdot 3p^2(1 - p) + 2 \cdot \left(\frac{1}{2}\right) \cdot 3p^2(1 - p)}_{2-\text{arrivals}} \tag{2.138}$$

$$= \underbrace{1 \cdot \left(\frac{1}{4}\right) \cdot p^3 + 2 \cdot \left(\frac{3}{4}\right) \cdot p^3}_{3-\text{arrivals}} \tag{2.139}$$

Now let's consider loss probability, the probability that a packet is lost, first from a birds eye view. Let's define $P_{\text{loss}|n}$ as the probability of loss if there are n arrivals. Also $P(n)$ is the probability of n arrivals. Naturally loss only can occur if there are two or three arrivals. Then

$$P_{\text{loss}} = P_{\text{loss}|2}P(2) + P_{\text{loss}|3}P(3) \tag{2.140}$$

The probability of n arrivals, $P(2)$ and $P(3)$, are the same as usual. In Figure 2.15, of the eight arriving packets, two are lost so that $P_{\text{loss}|2} = 0.25$. In Figure 2.16, of the 24 arriving packets, 10 are lost so that $P_{\text{loss}|3} = 5/12$. Thus

$$P_{\text{loss}} = \left(\frac{1}{4}\right) \cdot 3p^2(1-p) + \left(\frac{5}{12}\right) \cdot p^3 \tag{2.141}$$

For policy B, let's again look at a "tagged" arriving packet. This is the view from a packet that we have a seat on that we KNOW is arriving. Suppose first that the question is what is the loss probability (of *some* packet) given our tagged packet arrives. Then from Figure 2.15, half of the four boxes lead to loss, so $P_{\text{loss}|2} = 0.5$. For three arriving packets, from Figure 2.16, all of the eight boxes lead to loss, so $P_{\text{loss}|3} = 1.0$. However, since we "know" that our tagged packet is arriving, the probability of two arrivals is simply the probability of one more packet arriving. Similarly, for the same reason the probability of three arriving packets is the probability of two additional packets arriving. Then

$$P_{\text{loss}} = \frac{1}{2} \cdot 2p(1-p) + 1.0 \cdot p^2 \tag{2.142}$$

Now say that we are interested in the probability that OUR tagged packet is lost. For two packets arriving to the switching element, again half of the time loss occurs but since two packets are involved and the chance that the tagged packet is the one dropped is 0.5, so $P_{\text{loss}|2} = 0.5 \times 0.5 = 0.25$. If there are three arriving packets and all three packets prefer the same output (which occurs 2/8 of the time), the probability that the tagged packet is the one lost is 2/3 (i.e., two packets out of three are lost). If there are three arriving packets and two packets prefer the same output and the remaining packet prefers the other output (which occurs 6/8 of the time), the probability that the tagged packet is the one lost is 1/3 (one of three packets are lost). The probability of loss is modified in the equation above for

Thus

$$P_{\text{loss}} = \frac{1}{4} \cdot 2p(1-p) + 1.0 \cdot \left(\frac{2}{8} \cdot \frac{2}{3} + \frac{6}{8} \cdot \frac{1}{3}\right) \cdot p^2 \tag{2.143}$$

2.6.3 Networks

As indicated above, an engineer is interested not just in single switching elements but also in networks of switching elements and networks for packet distribution. In fact a complete study of networks of switching elements requires a knowledge of queueing theory (see chapter 3), so only some basic situations are covered below.

Knockout Switch

A great deal of effort since the 1980s has gone into designing high-speed packet switches (i.e., switching computers). In particular ATM switch designs have reached a sophisticated level. Here ATM is a high-speed packet switching technology optimized to carry different types of traffic such as voice, video, and data using 53 byte packets.

The knockout switch, designed by Yeh, Hluchyj, and Acampora (Yeh 87), is one such high-speed packet switch design. A system-level design is shown in Figure 2.17. The overall system has N inputs and N outputs. All N inputs go over buses to each of N bus interfaces (one for each output). A bus interface filters out packets that are destined for its output.

A key feature of the knockout switch is that each bus interface relays at most L packets arriving in a slot from the inputs to its output. Excess packets (beyond L) are dropped or erased. The L packets are sequentially fed to L shared buffers that lead to the bus interface's output. The term "knockout" is used because the interface contains a network (concentrator) implementing a sports knockout tournament strategy to randomly choose, using a VLSI circuit, which packets to send to the output if more than L packets arrive (Robertazzi 93, Saadawi 94, Yeh 87).

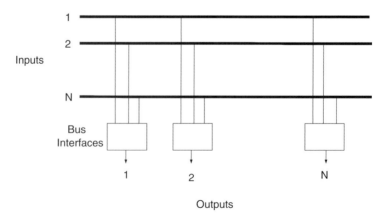

Fig. 2.17. Knockout switch system

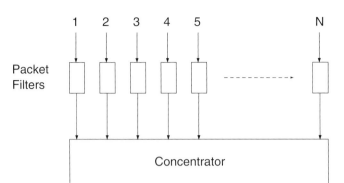

Fig. 2.18. Knockout switch bus interface for a single output

Assume that packet arrivals on each input line are aligned with arrivals on other input lines (i.e., arrivals are synchronous). It turns out that, if the traffic arrivals are uniform, it is statistically rare for more than L packets to arrive in a slot. To see this, let p be the independent Bernoulli arrival probability for each input. We assume uniform arrivals. That is, each input has the same arrival rate (p) and each packet is equally likely to go to any output. Then the probability that n packets arrive at a bus interface destined for that bus interface's output line is

$$P(n) = \binom{N}{n} \left(\frac{p}{N}\right) \left(1 - \frac{p}{N}\right)^{N-n} \quad n = 0, 1, 2, \ldots N \quad (2.144)$$

Here we use (p/N) rather than the p of the earlier binomial distributions in this chapter (section 2.3) as a packet arrives to an input with probability p but goes to a particular bus interface with uniform probability, $1/N$. Thus the relevant packet arrival probability at a bus interface for a particular output is p/N.

Let us find the (bird's eye view) probability of packet loss P_{loss}, that is, the probability a packet is dropped because the bus interface accepts some other L packets. The average number of packets lost by a bus interface is clearly the average number of packets arriving to the bus interface times the loss probability. Now Np packets arrive on average to the whole switch (N inputs each with arrival probability p). However $1/N$ of the Np packets goes to a single bus interface so that the average number of packets arriving to the interface unit is p. This is less than 1.0, but it is an average; more than one packet may arrive to the interface unit on occasion.

Thus the loss probability P_{loss} is

$$P_{\text{loss}} = \frac{1}{p} \times \overline{L} \qquad (2.145)$$

Here $\overline{L}$ is the average number of lost packets at an interface unit. Then

$$P_{\text{loss}} = \frac{1}{p} \left(\sum_{n=L+1}^{N} (n - L) \binom{N}{n} \left(\frac{p}{N}\right)^n \left(1 - \frac{p}{N}\right)^{N-n} \right) \qquad (2.146)$$

In this expression, as an example, if a bus interface accepts no more than two packets ($L = 2$) and five packets arrive to it ($n = 5$), the three lost packets ($n - L = 3$) weight the binomial probability of five arriving packets.

Figure 2.19 illustrates the loss probability versus different size switches for a $p = 0.9$ (90%) loading. Note that, if a bus interface accepts at most eight

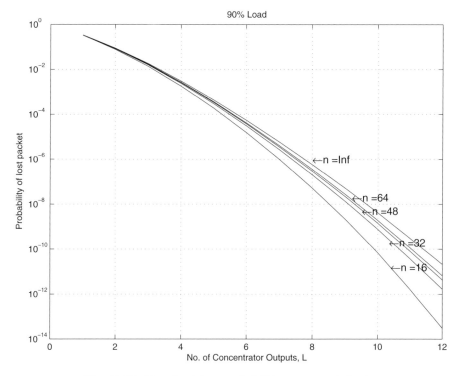

Fig. 2.19. Knockout loss probability versus switch size

packets, then the loss probability is less than one in a million for any size switch ($n > 8$).

This idea, that few packets are lost, breaks down if traffic is not uniform. If traffic is highly multiplexed, the uniformity assumption may be reasonable. If there is one or more "hot spot" output (outputs sinking an inordinately large number of packets), the loss probability will increase.

Crossbar Switch

One can use "interconnection networks" to do the switching in a high-speed packet switch. An interconnection network (Wu 84) is a network of relatively simple switching elements used to connect a set of inputs to a set of outputs. The connectivity may be circuit switched based or packet switched based. We have seen the evaluation of isolated switching elements in section 2.6.2. Networks of switching elements are optimized for such considerations as blocking probability, throughput, delay, and implementation complexity.

A fundamental interconnection network is the crossbar. This can be visualized as N horizontal wires from inputs and N vertical wires going to outputs, arranged in a grid where the wires normally do not touch (Figure 2.20). At each place where two wires cross in the grid, there is a "crosspoint" switch that, on command, electrically connects the wires. If one is implementing circuit switching, closing one/some of the switches connects desired inputs to desired outputs as in the figure. Note that, for point-to-point communication, only one crosspoint per row and column may be closed at one time.

In using a crossbar interconnection network for packet switching, crossbar switches are closed long enough to forward a packet from input to output. Packet transmissions through the crossbar may be synchronized. That is, switches close in synchronism for one slot time to transmit fixed length

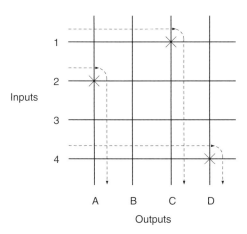

Fig. 2.20. Crossbar switch

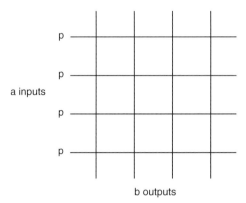

a inputs

b outputs

Fig. 2.21. Crossbar packet switch

packets (as in ATM). Buffers may be placed at each input, each output, or both. One places buffers at inputs to store arriving packets that can't immediately get through the switch because of previously stored packets (there is queueing). Buffers may be placed at outputs if more packets may occasionally arrive during a slot for a given output than the output line (from the buffer) can transmit in one slot. Performance evaluations and design trade-offs for buffered crossbar interconnection networks appear in (Hui 87, Karol 87, Oie 89, Robertazzi 93a, Robertazzi 00, Yoon 90). Major design trade-off issues are implementation complexity/feasibility and throughput.

In this subsection we consider a crossbar without buffers, as in Figure 2.21. As described in Patel 81, let fixed length packets fit exactly in one time slot. In each slot a packet may arrive to one of a inputs with independent probability p (and not arrive with independent probability $1 - p$). There are b outputs. Arrivals and switch closings are synchronized (occur together in each slot). Even if more than one packet arrives for a specific output, only one (randomly chosen) packet is actually forwarded to the output. Others are dropped. This is a consequence of the fact that only one switch on a vertical wire leading to an output may close during each slot.

The problem here is to find the throughput of the crossbar under a uniform load. The throughput is less than 1.0 (100%) as packets are dropped.

The answer to this problem involves a binomial distribution. In a manner similar to the previous knockout switch problem, the probability an input sends a packet to a specific output is p/b under this uniform load. That is, a packet would arrive on a given input with probability p and go to a specific output with probability $1/b$.

The probability that n packets from the inputs attempt to go to a specific output in a slot is

$$P(n) = \binom{a}{n} \left(\frac{p}{b}\right)^n \left(1 - \frac{p}{b}\right)^{a-n} \qquad n = 0, 1, 2 \ldots a \qquad (2.147)$$

The average throughput for one output line is the probability that one or more packets try to go to that output

$$\overline{\text{Throughput}}_{\text{output}} = P(n \geq 1) = 1 - P(0) \tag{2.148}$$

$$\overline{\text{Throughput}}_{\text{output}} = 1 - \left(1 - \frac{p}{b}\right)^a \tag{2.149}$$

The total throughput for the whole switch is b times the throughput of one output line or

$$\overline{\text{Throughput}}_{\text{output}} = b\left(1 - \left(1 - \frac{p}{b}\right)^a\right) \tag{2.150}$$

For a very large switch $(a, b \to \infty)$ under a heavy load $(p = 1)$ since

$$\lim_{x \to \infty} (1 - 1/x)^x = \frac{1}{e} \tag{2.151}$$

One has

$$\lim_{a,b \to \infty} b\left(1 - \left(1 - \frac{p}{b}\right)^a\right) = b\left(1 - \frac{1}{e}\right) \tag{2.152}$$

$$\boxed{\lim_{a,b \to \infty} \overline{\text{Throughput}}_{\text{output}} = 0.632\,b} \tag{2.153}$$

So under a uniform heavy load, a very large switch has a 63.2% average throughput. That is, 63.2% of the packets get through the crossbar. If the load is not uniform, there will be deviation from this value.

Multiple Bus System

Using a bus to interconnect stations is called a shared media connection. The simple discrete time Ethernet model of section 2.4.2 involves N stations sharing a bus. What if there are multiple buses, and each station is attached to several of them?

Suppose one has N stations, M buses, and R connections to different buses per station $(R \leq M)$. Figures 2.22 and 2.23 show two possible architectures.

From Figure 2.22 it is clear that only certain combinations of N, M, and R lead to symmetric networks. An important parameter of this problem is the number of connections per bus or NR/M. This is the number of system-wide connections divided by the number of buses. For a symmetric system it must be an integer.

Let p be the probability a station attempts to transmit a packet on a given bus in a slot. Here we allow one station to attempt to transmit multiple (different) packets on more than one bus in a slot. As in the earlier discrete

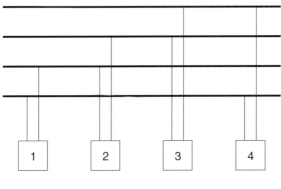

N=4, M=4, R=2

Fig. 2.22. A multibus architecture

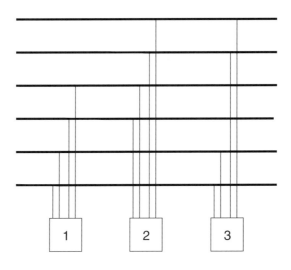

N=3, M=6, R=4

Fig. 2.23. Another multibus architecture

time Ethernet example, only if one station attempts to transmit on a specific bus is there a successful transmission (more than one station attempting to transmit on a specific bus causes useless collisions).

For the throughput of one bus, one has a binomial distribution

$$\overline{\text{Throughput}}_{\text{bus}} = \binom{NR/M}{1} p(1-p)^{\frac{NR}{M}-1} \tag{2.154}$$

For the whole system of M buses, the throughput is M times as big

$$\overline{\text{Throughput}}_{\text{system}} = M \left(\frac{NR}{M} \right) p(1-p)^{\frac{NR}{M}-1} \qquad (2.155)$$

$$= NRp(1-p)^{\frac{NR}{M}-1} \qquad (2.156)$$

If p is too small, many slots on buses will be idle. If p is too large, there will be many collisions. Both situations reduce throughput. The optimal choice of p can be found by taking the derivative of mean system throughput with respect to p.

Since, in calculus, for two functions $f(p)$ and $g(p)$

$$\frac{df(p)g(p)}{dp} = f(p)\frac{dg(p)}{dp} + g(p)\frac{df(p)}{dp} \qquad (2.157)$$

One has

$$\frac{d}{dp}\overline{\text{Throughput}}_{\text{system}} =$$

$$NR(1-p)^{\frac{NR}{M}-1} - NRp(1-p)^{\frac{NR}{M}-2} \left(\frac{NR}{M} - 1 \right) = 0 \qquad (2.158)$$

With algebra one can show

$$\boxed{p_{\text{optimal}} = M/NR} \qquad (2.159)$$

This makes intuitive sense. If p is M/NR and there are NR/M connections per bus, the total average offered load in a bus is $(M/NR) \times (NR/M) = 1.0$. Thus on average one station transmits on the bus at a time. This is an average. Sometimes, no stations or more than one station attempts to transmit on a bus. But with offered load averaging to one station access per slot per bus, throughput is maximized as this is the condition under which successful communication takes place.

This problem formulation is quite general. As in section 2.4.2, if there is one bus, $M = 1$ and $R = 1$ so that mean system throughput is

$$\overline{\text{Throughput}}_{\text{system}} = Np(1-p)^{N-1} \qquad (2.160)$$

If there are M buses and each station connects to every bus, $M = R$ and

$$\overline{\text{Throughput}}_{\text{system}} = MNp(1-p)^{N-1} \qquad (2.161)$$

This M-fold increase in throughput is consistent with the previous equation.

2.7 Conclusion

Since its inception several hundred years ago, probability has been a powerful tool to evaluate uncertainty in various systems. This chapter has introduced basic concepts such as the Bernoulli process and related distributions in the context of important networking concepts and applications.

Once one has developed skills in applying probability to technological problems, the skills can be applied to new technologies as they arise. Even if one never evaluates a high-speed packet-switching architecture one may work with ad hoc radio networks, TCP/IP versions, digital communication systems, or other systems and be glad one has a bag of probability tools.

2.8 Problems

1. Suppose a simple Markovian queue has state-dependent arrival and service rates. That is, the arrival and service mean rates are functions $\lambda(n)$ and $\mu(n)$ (respectively) of the number of customers in the queue just prior to an arrival or departure. Rewrite the development of equations (2.15) to (2.20) in these terms.
2. Suppose a telephone switch accepts 400 calls/second during a certain time period, according to a Poisson process. Using the Poisson distribution find the probability of zero, one, or two calls during a time interval of $1/400$ seconds and during a $1/1000$-second interval. Do the numbers make intuitive sense?
3. If 400 calls/second arrive to a telephone switch, according to a Poisson process, what is the average number of calls arriving in $1/400$ second? In $1/4$ second? In 1 second?
4. Software in a telephone switch can process 400 calls/second. If the call arrival rate is tripled, how many more calls arrive in 0.25 seconds?
5. Show that equation (2.34) is correct for its application.
6. The geometric distribution of equation (2.36) decreases as i increases. Intuitively why?
7. Every time a telephone truck leaves the depot, there is a $1/500$ probability that it breaks down. How many days from today, on average, can the first breakdown be expected? Hint: What statistical process and distribution is implied? Why might this not be a good reliability model?
8. A packet train is a series of consecutive time slots with packets. The probability of a packet arrival in a slot is p. Given that one packet arrives and starts the train, what is the average length of the train? Hint: Think of the events that end the packet train.
9. What value of p maximizes the variance in the number of arrivals described by a Bernoulli process? Why?
10. Prove that the mean (average) number of arrivals for a Bernoulli process with packet arrival probability p for N total slots is Np.

11. Packets arrive to an input line of a packet switch according to a Bernoulli process. The packet arrival probability is $p = 0.2$. What is the probability of five arrivals in ten slots in any pattern of arrivals?

12. In the previous problem, the probability of three arrivals in ten slots, when $p = 0.5$ is the same as the probability of seven arrivals, in ten slots with $p = 0.5$. Why?

13. Consider three telephone circuits from the United States to Europe belonging to a company. Measurements reveal that during the business day each circuit is busy 20 minutes out of an hour, on average. Most calls are short (2 or 3 minutes or less).

   (a) Let p be the independent probability that a single circuit is busy. Find a numerical value of p and show how you arrived at it.

   (b) Write an expression for the probability a given call goes through (finds a free circuit).

   (c) Write an expression for the probability that, at a given instant, *exactly* one circuit is free.

   (d) Write an expression for the average number of busy channels. Hint: Write it as a weighted sum.

14. A catalog company has its own computerized consumer order system. There are three regional centers. Each regional center (New York, Dallas, and San Francisco) has two independent computers for reliability. Let p be the independent probability that a single computer is down. For the entire network to be considered to be functional, at least one computer must be up in each regional center. Find an expression for the probability that the entire network is functional.

15. For a Bernoulli arrival process, find the smallest value of p so that the probability of exactly four packets arriving in ten slots is at least 0.6.

16. Show that the variance of a binomial distribution is:

$$\sigma^2 = Np(1 - p) \tag{2.162}$$

17. Packets arrive according to a Bernoulli process to an input line of a packet switch. Find the probability that the fifth arrival occurs in exactly ten slots, if $p = 0.35$.

18. In the previous problem, what is the average number of slots holding 5 arrivals if $p = 0.35$.

19. Plot the throughput of the discrete time simple Ethernet model [equation (2.54)] versus offered load, p. Let $N = 10$. Also plot the probability of collision versus p [equation (2.56)]. Comment on the shape of the curves.

20. For the simple Ethernet model of section 2.4.2 show that the optimal (throughput maximizing) choice of p, the station transmission probability is $1/N$. See equation (2.54).

21. Referring to the Ethernet design equation, as the probability that only a single station acquires a channel increases, the average number of slots

per contention interval decreases [see equation (2.64)). Intuitively, why is this so?

22. Find the Ethernet utilization (efficiency) for a 100 Mbps network of size 50 meters and a 512 bit minimum frame size.

23. What is more realistic for designing a 10 Gbps Ethernet: reducing network physical size by a factor of 10 or increasing the minimum frame size by a factor of 8 to 10? Refer to equation (2.70) and the following discussion.

24. Explain why the window of vulnerability for pure Aloha is equal to 2T [see discussion above equation (2.77)].

25. Show that throughput is maximized at 18.4% for pure Aloha at $G = 0.5$ and at 36.8% for slotted Aloha at $G = 1.0$.

26. Which system, slotted or pure Aloha, has a larger mean delay as G becomes larger? Intuitively why?

27. Verify equation (2.96) using calculus.

28. Verify that equation (2.99) is equivalent to equation (2.98) using calculus.

29. Find the load in link x in the capacity calculation section for a binary tree (section 2.5.3) if there are two, five, and ten levels.

30. Consider a tree as in the capacity calculation subsection of section 2.5.3 and Figure 2.11. Calculate the capacity needed at an arbitrary link in the tree using the same uniform loading assumption as in the capacity calculation section.

31. Consider a network load intensity problem, similar to section 2.5.4, except the network area has a square shape. Sketch an intuitive representation of the network load intensity. Explain your answer.

32. Find the expression for throughput in the multiple bus problem of section 2.6.3 if a station transmits in a slot with probability p to a specific randomly chosen bus (with probability $1/M$). That is, a station attempts to transmit in only one bus in a slot.

33. Consider a diamond-shaped network of four links and four nodes. The left corner node is A, the top corner node is B, the bottom corner node is C, and the right corner node is D. Sketch a diagram. Links AB, BD, AC, and CD carry at most one circuit each. Here p is the independent probability that a link is available (idle).
(a) Find the probability that the upper path from A to D (ABD) is available (both links available). Call this probability q.
(b) Find the probability at least one of the paths from A to D (upper and lower) is available.

34. Consider three nodes, A, B, and C. You can draw them horizontally. There are three links from A to B and three links from B to C. Each link carries at most one circuit. At any time p is the measured probability that each link is in use (and not accepting further circuits).
(a) Find an expression for the average number of busy links (actually a number between 0 and 6). There is both a simple and a more elaborate answer.

(b) Find an expression for the probability that at least one idle path (two consecutive idle links) exists from nodes A to C.

35. Consider now a model like the last problem but with four nodes, A, B, C, and D. There are three links each between A and B, B and C, and C and D. Again p is the independent probability that each of the (nine) links is busy (each link holds at most one circuit). We are interested in circuits (paths) from A to D. A circuit for a call from A to D uses any available link from A to B, B to C, and C to D.

    (a) Find the probability that a call from A to D is "blocked" (i.e., there is no available path from A to D).

    (b) Find the probability that at least two (i.e., two or three) paths from A to D are available.

36. Consider a small business with ten phones connected to two outgoing lines through a PBX (private branch exhange switch). A phone seeks an outside line with independent probability p (and is idle with probability $1 - p$). Only two phones, at most, can utilize the two outside lines at a time (one phone per outside line).

    (a) Find an expression for the probability that n phones wish to seek an outside line.

    (b) Find the probability of "blocking" occurs (one or more calls can't get through when they want to). This should be a function of p.

37. Consider a switching element with two inputs. Time is slotted, and the independent probability of an arrival in a slot at an input is p. The probability of no arrival at an input in a slot is $1 - p$.

    Find an expression for the probability of three or more packets arriving in two consecutive slots over both inputs.

38. Consider a 2 input, 2 output switching element. Time is slotted. Arrivals occur to each input, each slot, with independent Bernoulli probability p. If only one packet arrives to the inputs, it is randomly routed to one of the outputs (with probability 0.5 of going to a specific output). If two packets arrive to the inputs (one on each input), each goes to a separate output. For the purposes of this problem it doesn't matter which packet goes to which output in the latter case of two arriving packets.

    (a) Find the probability of there being a packet on a specific (given) output in a slot. Show how you arrived at your answer.

    (b) Find an expression for the expected (average) number of packets at the outputs in a slot.

39. Consider a 3 input, 2 output switching element. One output has a connection that feeds its packets back to one of the inputs so that there are two external inputs and one external output. The probability of a packet arrival at an external input in a slot is p (and the probability of no arrival is $1 - p$).

    The switching element policy is that one of three input packets is selected randomly to go to the external output. If two or more packets wish to go to the output, one is selected for the external output, one is selected for the feedback output, and any remaining packet is erased (dropped).

(a) Find an expression for q, the probability that a packet is fed back in a slot. This will be an implicit equation where q is a function for q.

(b) Solve part (a) for q (i.e., find an explicit equation).

40. Consider a multicasting network of 1 input, 2 output switching elements. The network has a single input to the first switching element (stage). This switching element's two outputs each go to another (of two) switching element (stage 2). The stage 2 switching element outputs each go to (one of four) switching elements (stage 4). Thus there are seven elements, one network input and eight network outputs.

    For each switching element, if there is an input packet, a copy appears on each of the two element outputs with independent probability c.

    Packet arrivals to the network input occur in a slot with independent probability p (with no arrival with probability $1 - p$).

    (a) Find an expression for the probability that a copy appears at a network system output.

    (b) Find an expression for the average number of copies produced at the network outputs in one slot.

41. Sketch a 4 input, 2 output switching element. Time is slotted and synchronized at both inputs and outputs. One slot holds at most one packet. Packet arrival processes to each input are Bernoulli (with arrival probability p for a packet in each slot and non-arrival probability $1 - p$ in each slot).

    (a) Determine the probabilities of zero to four arrivals in a slot.

    (b) Determine the probability of at least one arrival in a slot, simply.

42. Sketch a packet switching element with 4 inputs and 2 outputs. Time is slotted and synchronized at both inputs and outputs. One slot holds at most one packet. Packet arrival processes to each input are Bernoulli (with independent probability p for a packet in each slot and independent non-arrival probability $1 - p$ in each slot).

    If one or two packets arrive in a slot, they are transmitted on the output(s). If three or four packets arrive, two packets are randomly chosen to be transmitted on the outputs (one on each at most) and the remaining packet(s) are lost (erased). Packets are assigned to outputs randomly.

    (a) Determine from a "bird's eye" view the probability that a packet is dropped.

    (b) Determine the probability a given (tagged) arriving packet is dropped.

    (c) Determine the mean throughput of the switching element. This is the mean number of packets transmitted on the output links.

43. Sketch a switching element with N inputs and 3 outputs. Time is slotted and synchronized at both inputs and outputs. One slot holds at most one packet. Packet arrival processes to each input are Bernoulli (with arrival probability p for a packet in each slot and non-arrival probability $1 - p$ in each slot).

If one, two, or three packets arrive to the inputs in a slot, the packets are transmitted on the outputs. If more than three packets arrive at the element input, in a slot, three are randomly chosen to be transmitted on the outputs and the remainder are lost. Packets are assigned to outputs randomly.

(a) Determine the mean switching element throughput (as a function of p and N).

(b) Determine the mean number of dropped packets during a slot as a function of p and N.

44. Sketch a network of four switching elements. All switching elements, A, B, C, and D, have three inputs and one output. The inputs of element D are A, B, and C's outputs. Thus, the overall system has nine inputs and one output. Time is slotted and synchronized at both inputs and outputs. One slot holds at most one packet. Packet arrival processes to each input are Bernoulli (with independent probability p for a packet arrival in each slot and independent non-arrival probability $1 - p$ in each slot).

For each switching element, if one packet arrives at its inputs in a slot, it is transmitted on the single element output. If more than one packet arrives to element inputs in a slot, one packet is randomly chosen to be transmitted through the output and the remaining packets are lost.

(a) Determine the probability q that a packet is at the output of either elements A, B, or C in a slot.

(b) Find the mean throughput of the system.

45. Find and sketch two different attachment patterns from the ones illustrated for the multiple bus system in section 2.6.3.

46. For the multiple bus system of section 2.6.3, if p is 10% larger (smaller) than the optimal p, find the percentage change in throughput (let there be three stations, six buses, and four bus attachments per station).

47. **Computer Project**: Plot the Poisson distribution (section 2.2) for $n = 0, 1, 2, 3$. Plot the probability $P_n(t)$ versus t when $\lambda(t) = 1$.

48. **Computer Project**: Plot the knockout switch loss probability [equation (2.146)] on a log scale versus the number of concentrator outputs (1 through 12) for $N = 16, 32, 64$ and N very large with a 90% load ($p = 0.9$).

49. **Computer Project**: Plot the crossbar output throughput of equation (2.150) as a function of p for $a = b$ from 2 through 30 in steps of 2.

50. **Computer Project**: Plot the system throughput of a multiple bus system [equation (2.156)] versus p, where $N = 12$, $R = 3$, and $M = 2, 3, 4, 6, 9, 12, 16$.

3
Queueing Models

3.1 Introduction

The concept of using mathematical models to evaluate the carrying capacity of communications devices began in the early years of the telephone industry. Around 20 years into the twentieth century, the Danish mathematician A. K. Erlang applied the theory of Markov models developed by the Russian mathematician A. A. Markov 10 years earlier to predicting the capacity of telephone systems. Erlang's brainchild went on to be called queueing theory.

A "queue" is the British word for a waiting line. Queueing theory is the study of the statistics of things waiting in lines. Calls may wait at an exchange, jobs may wait for access to a processor, packets of data to be networked may wait in an output line buffer, planes may wait in an holding pattern, and a customer may wait at a supermarket checkout counter. For over 80 years queueing theory has been used to study these and similar situations.

This chapter introduces basic queueing theory using Markov models. In fact advanced research in queueing theory involves non-Markov models (such as self-similar traffic) but Markov models remain the foundation of queueing theory and will allow the reader of this introductory text entry to the world of queues.

In section 3.2 of this chapter, two basic queueing models are examined. One is the continuous time M/M/1 model of Erlang. The second model is the more recent discrete time GEOM/GEOM/1 model. Section 3.3 discusses some important specific single queue models. Common performance measures are described in section 3.4. Networks of queues and an associated computational algorithm are presented in sections 3.5 and 3.6, respectively.

The special cases of queues with negative customers, recursive solutions for the equilibrium state probabilities of non-product form networks, and stochastic Petri nets are discussed in sections 3.7, 3.8, and 3.9 respectively. Finally, section 3.10 covers numerical and simulation solution techniques for models.

3.2 Single Queue Models

3.2.1 M/M/1 Queue

The M/M/1 is the simplest of all Markov-based queueing models. In the standard queueing model notation the first position in the descriptor indicates the arrival process. The "M" here represents a Poisson arrival process (see section 2.2), with the M representing the memoryless (Markovian) nature of the Poisson process. The second position in the M/M/1 descriptor indicates that the time spent in the server by a customer follows a negative exponential distribution

$$\mu e^{-\mu t} \tag{3.1}$$

Here μ is the mean service rate. This is the only continuous time distribution that is memoryless in nature, allowing for the "M" in the descriptor's second position. That is, if a customer has been in the server for x seconds, the distribution of time until it completes service is also negative exponentially distributed as above. An intuitive explanation of the memoryless property of the negative exponential distribution is that the negative exponential distribution is a continuous time analog of a random coin flip, where the history of past flips has no influence on the future. In fact the negative exponential distribution is a good model for the duration of voice telephone calls. Finally, the last position in the M/M/1 descriptor indicates the number of servers fed by the waiting line. A server is where the actual processing takes place, as at the checkout counter at a supermarket. The usual assumption is a server holds at most one customer at a time. Here "queue" may refer to the waiting line only or the whole system of waiting line and server(s). This is usually clear from the context.

Queues and queueing networks have a schematic language that gives a higher level application view of the underlying Markov chain. Figure 3.1 shows a single-server queue. The open box is the waiting line and the circle represents the server. Figure 3.2 illustrates a four-server queue, as in some banks, with one common waiting line and four tellers.

The schematic queue of Figure 3.1, with Markovian statistics (M/M/1) gives rise to the state transition diagram of Figure 3.3. A state transition

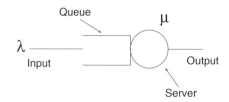

Fig. 3.1. Single-server queue schematic

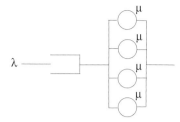

Fig. 3.2. Four-server queue schematic

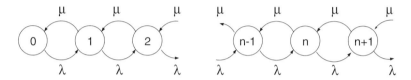

Fig. 3.3. M/M/1 queue state transition diagram

diagram is a stochastic finite state machine. The circles are states $(0, 1, 2 \ldots$ customers in the entire queueing system). The transitions correspond to arrivals of customers (which change the state from n to $n + 1$ with mean rate λ) and departures of customers from the server (which change the state from n to $n - 1$ with mean rate μ).

Because the transitions only increment or decrement the number of customers in the queueing model by one, this type of model is called a birth death population process in statistics. More elaborate models allow "batch" arrivals or departures, that is, multiple arrivals or departures at the same time instant. Naturally the corresponding transitions may jump several states.

This discussion raises an interesting point. In such a continuous time model, only one transition is made (in zero time) at a time. That is, one never has an arrival or departure at the same time instant. Of course in a continuous time model, time is fine enough that this is possible (if one has an arrival at $T = 2.72316$ seconds, a departure may be at $T = 2.72316001$ seconds).

The state transition diagram of Figure 3.3 is like an electric circuit in the sense that something "flows" through the graph. That something is probability flux. The flux flowing over a transition at time t is simply equal to the product of the transition rate of the transition and the state probability at time t at the state at which the transition originates. Note that Markov chains are different from electric circuits in that for electric circuits flow direction is not preset (it depends on a voltage difference across an element) and a battery or voltage source guarantees nonzero flows in a circuit. In a Markov chain the

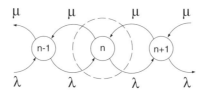

Fig. 3.4. Global balance boundary about state n

normalization equation, which holds that the sum of state probabilities is one, guarantees nonzero flows.

Probability flux, intuitively, has units of [events/second] and represents the mean number of times per second a transition is transited. A Markov chain may be viewed as a "lattice" on which the system state performs a random walk. Random walks are widely studied random processes (Papoulis 02). Under the conditions of a continuous time Markov chain random walk, the system state stays a negative exponential distributed random time in a state with a mean time equal to the inverse of the sum of the outgoing transition rates. At the end of this time the system state enters a neighboring state on a random basis. In Figure 3.3 the system state will leave an interior state n for state $n + 1$ with probability $\lambda/(\lambda + \mu)$ and leave state n for state $n - 1$ with probability $\mu/(\lambda + \mu)$. The pattern should be clear. The probability that the system state leaves a state on a particular transition is equal to that transition's transition rate divided by the sum of outgoing transition rates from the state.

One can set up a series of differential equations to solve for the state probabilities for the M/M/1 model (Kleinrock 75, Robertazzi 00). An alternative approach that will be pursued here is to use the concept of global balance. Global balance, which is analogous to Kirchoff's current law for electric circuits, holds that the difference between the total flow of probability flux into any state and the total flow of probability flux out of the state at any point in time is the change of probability at the state at that time.

Referring to Figure 3.4 the total flow into state n is

$$\lambda P_{n-1}(t) + \mu P_{n+1}(t) \tag{3.2}$$

Here $P_n(t)$ is the nth state probability at time t. The first term represents the probability flux from state $n - 1$ to n, and the second term represents the probability flux from state $n + 1$ to n. Also the total flow out of state n to states $n - 1$ and $n + 1$ is

$$-(\lambda + \mu)P_n(t) \tag{3.3}$$

Note here that a negative sign here indicates outward flow and that a positive sign indicates inward flow. As discussed, the difference between the above two quantities represents the change in probability at state n as a function of time. Thus

$$\frac{dP_n(t)}{dt} = -(\lambda + \mu)P_n(t) + \lambda P_{n-1}(t) + \mu P_{n+1}(t) \qquad n = 1, 2 \ldots \quad (3.4)$$

Using a similar argument, the boundary state $(n = 0)$ can be modeled by

$$\frac{dP_0(t)}{dt} = -\lambda P_0(t) + \mu P_1(t) \qquad (3.5)$$

Naturally normalization holds that

$$P_0(t) + P_1(t) + P_2(t) + \cdots + P_n(t) + \cdots = 1 \qquad (3.6)$$

These equations represent a family of differential equations to model the time-varying behavior of the M/M/1 queue. In fact such transient solutions can be quite complex, even for the apparently simple M/M/1 queue. For instance, if one starts with i customers in the queue at time $t = 0$, the probability that there are n customers in the queue at time t is a function of modified Bessel functions of the first kind (Robertazzi 00).

The differential equation model of the M/M/1 queue allows transient modeling. However an important type of system behavior with a far simpler solution is steady state (or equilibrium) operation. In other words, if one waits long enough for transient effects to settle out while system parameters (like arrival and service rate) are constant, then system metrics approach a constant and the system exhibits a steady state type of behavior. The M/M/1 queueing system is still a stochastic system, but performance metrics like mean throughput, mean delay, and utilization approach a constant. In this mode of operation, the state probabilities are constant so

$$\frac{dP_n(t)}{dt} = 0 \qquad n = 0, 1, 2 \ldots \qquad (3.7)$$

This leads the previous family of differential equations to become a family of linear equations

$$-(\lambda + \mu)p_n + \lambda p_{n-1} + \mu p_{n+1} = 0 \qquad n = 1, 2 \ldots \qquad (3.8)$$

$$-\lambda p_0 + \mu p_1 = 0 \qquad (3.9)$$

This last equation is a boundary equation for the state $n = 0$. In these linear equations, the time argument for state probability has been deleted as the equilibrium probabilities are constant. These equations can also be simply found by writing a global balance equation for each state.

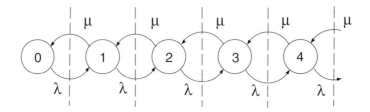

Fig. 3.5. Local balance boundaries between adjacent states

It turns out that, to solve any Markov chain with N states, one writes a global balance equation for each state. Any one of the N equations is redundant (contributes no unique information) and can be replaced by the normalization equation

$$p_0 + p_1 + p_2 + \cdots + p_{N-1} + p_N = 1 \qquad (3.10)$$

These N equations and N unknowns can be solved for the equilibrium state probabilities using a standard linear equation solver algorithm. This is of eminent interest because useful performance metrics such as mean throughput, delay, and utilization are simple functions of the state probabilities.

The problem one runs into in trying to use this approach for performance evaluation is that even modestly sized models can have a vast number of states. Moreover a direct linear equation solution has a computational complexity that is proportional to the cube of the number of states. This further compounds the problem. For instance a closed (sealed) queueing network of 9 queues and 60 customers has over 7 billion states! One can see that modeling the Internet in detail, with this approach at least, with its thousands of nodes and millions upon millions of packets, is out of the question.

However a single M/M/1 queue with its one-dimensional state transition diagram is quite easy to solve, even with an infinite number of states. The way this can be accomplished is to use a different type of balancing called local balance. For the one-dimensional Markov chain, one sets up boundaries between each adjacent pair of states (Figure 3.5). In equilibrium, probability (flux) should not build up or be depleted at each state. Thus one can equate the flow of probability flux across each boundary from left to right to the flow of probability flux across the boundary from right to left to obtain the following family of equations:

$$\lambda p_0 = \mu p_1 \qquad (3.11)$$

$$\lambda p_1 = \mu p_2 \qquad (3.12)$$

$$\lambda p_2 = \mu p_3 \qquad (3.13)$$

.

.

$$\lambda p_{n-1} = \mu p_n \tag{3.14}$$

A bit of algebra results in

$$p_1 = \frac{\lambda}{\mu} p_0 \tag{3.15}$$

$$p_2 = \frac{\lambda}{\mu} p_1 \tag{3.16}$$

$$p_3 = \frac{\lambda}{\mu} p_2 \tag{3.17}$$

.

.

$$p_n = \frac{\lambda}{\mu} p_{n-1} \tag{3.18}$$

.

Suppose now that we chain the equations together. That is, substitute the first into the second, this into the third, and so on to obtain

$$p_1 = \left(\frac{\lambda}{\mu}\right) p_0 \tag{3.19}$$

$$p_2 = \left(\frac{\lambda}{\mu}\right)^2 p_0 \tag{3.20}$$

$$p_3 = \left(\frac{\lambda}{\mu}\right)^3 p_0 \tag{3.21}$$

.

.

$$p_n = \left(\frac{\lambda}{\mu}\right)^n p_0 \tag{3.22}$$

.

Now all the equilibrium probabilities are functions of λ and μ (which are known) and p_0 (which is unknown but soon won't be).

To solve for p_0, one can write the normalization equation

$$p_0 + p_1 + p_2 + \cdots + p_n + \cdots = 1 \tag{3.23}$$

Substituting the previous equations for the state probabilities

$$p_0 + \frac{\lambda}{\mu}p_0 + \left(\frac{\lambda}{\mu}\right)^2 p_0 + \cdots + \left(\frac{\lambda}{\mu}\right)^n p_0 + \cdots = 1 \tag{3.24}$$

Factoring out p_0 yields

$$p_0 \left(1 + \frac{\lambda}{\mu} + \left(\frac{\lambda}{\mu}\right)^2 + \cdots + \left(\frac{\lambda}{\mu}\right)^n + \cdots \right) = 1 \tag{3.25}$$

or

$$p_0 \left(\sum_{n=0}^{\infty} \left(\frac{\lambda}{\mu}\right)^n \right) = 1 \tag{3.26}$$

$$p_0 = 1 / \left(\sum_{n=0}^{\infty} \left(\frac{\lambda}{\mu}\right)^n \right) \tag{3.27}$$

This can be simplified further. Let $\rho = \lambda/\mu$. From the appendix, we know that

$$\sum_{n=0}^{\infty} \rho^n = \frac{1}{1-\rho} \tag{3.28}$$

Here $0 \leq \rho \leq 1$. Then

$$p_0 = 1 / \left(\sum_{n=0}^{\infty} \left(\frac{\lambda}{\mu}\right)^n \right) = 1 / \left(\sum_{n=0}^{\infty} \rho^n \right) = 1 - \rho \tag{3.29}$$

Also

$$p_n = \left(\frac{\lambda}{\mu}\right)^n p_0 \tag{3.30}$$

$$\boxed{p_n = \rho^n (1 - \rho) \qquad n = 0, 1, 2, ..} \tag{3.31}$$

This is an extremely simple and elegant expression for the equilibrium state probability of the M/M/1 queue. The particularly simple expression, even though there are an infinite number of states, is partly because there is a simple expression for the associated infinite summation.

We can deduce an important fact about the M/M/1 equilibrium state probability distribution from the above expression. The term $(1 - \rho)$ is a constant, and the term ρ^n decreases as N increases (since $\rho < 1$). Thus, the probability that the M/M/1 queue is in a state n (holds n customers) decreases as n increases. Thus, for example, if λ equals 1 customer/second and μ equals 3 customers/second, the probability that the queue holds 100 customers is $(1/3)^{100}$ or 1.9×10^{-48} times smaller than the probability that there is one customer. With the geometric-like decrease in state probability as N increases, an M/M/1 queue only has a relatively small number of customers most of the time.

Another important fact about the M/M/1 queue is that, for an infinite-sized buffer model as we have here, the arrival rate must be less than the service rate ($\lambda < \mu$ or $\rho < 1$) Otherwise customers would arrive faster than the queue could dispose of them and the queue size would increase without limit in an unstable mode of operation. The condition for infinite buffer M/M/1 queue stability is simply $\rho < 1$.

If the arrival process to an M/M/1 is a Poisson process, what can be said of the departure process? In fact a theorem due to Burke (Burke) shows it is also Poisson. Despite the appealing symmetry, this is not at all obvious. Recall that a process of rate λ is Poisson if and only if the interarrival (interevent) times are independent negatively distributed exponential random variables. If a queue has customers, the interdeparture time is indeed negative exponentially distributed, although with service rate μ. That is, the time between departures is simply the service completion time. But sometimes the queue is empty!

In that case the time between a departure that empties the queue and the next departure is the sum of two random variables, the negative exponential arrival time (with rate λ) and the negative exponential service time of that first arriving customer (with rate μ). The distribution of the sum of two such random variables is not negative exponential!

But it turns out that, in the totality of the output process, the statistics are Poisson. Although a proof of this is beyond the scope of this book, the concept of "reversibility" can be used in this effort. A reversible process is one where the statistics are the same whether time flows forward or backward. To the author's knowledge, no one has used this to create a time machine, but certainly such an effort would involve a great many states!

A very general result that applies to many queueing systems, including the M/M/1 queue is Little's Law. It is usually written as:

$$L = \lambda W \qquad (3.32)$$

Here L is the average queue size (length), λ is the mean arrival rate, and W is the mean delay a customer experiences in moving through a queue. As an example, if 40 cars an hour arrive to a car wash and it takes the average car 6 minutes (0.10 hours) to pass through the car wash, there is an average of about 4 cars (40×0.10) at the car wash at any one time.

Little's Law will even apply to a queueing system where the interarrival and service times follow specified arbitrary (general) distributions. Such a queue with one server is called a G/G/1 queue.

3.2.2 Geom/Geom/1 Queue

The Geom/Geom/1 queue is a discrete time queue (Woodward 94, Robertazzi 00). That is, time is slotted and in each slot there is zero or one arrivals to the queue and zero or one departures. Arrivals are Bernoulli so p is the independent probability of an arrival in a slot to the queue and $1 - p$ is the probability of no arrivals. We'll let s be the independent probability that a packet in the server departs in a slot. Thus, even though a packet may be in the server, it may not depart in a given slot.

The time to an arrival and the time for a departure to occur, measured in slots, are naturally given by a geometric distribution (see section 2.3). The geometric distribution is the discrete time analog of the continuous time negative exponential distribution. In fact the geometric distribution is the only memoryless discrete time distribution, just as the negative exponential distribution is the only memoryless continuous time distribution. If one looks at the underlying Bernoulli process, one can see that, just as in a series of coin flips, the time to the next head (arrival) is a memoryless quantity.

The Markov chain of the Geom/Geom/1 queue is shown in Figure 3.6. Once again, we have equilibrium state probabilities $P_0, P_1, P_2 \ldots P_n \ldots$. Here capital P is used for the equilibrium state probabilities to distinguish them from the arrival probability p. Now, though, instead of transition rates we have transition probabilities (the probability that a transition from state i to state j is made from one slot to the next is $t_{i,j}$).

The probability of going from state n to state $n+1$, thereby increasing the number of packets in the queue by one, is the probability of an arrival and no service completion or $p(1 - s)$. The probability of going from state n to state $n-1$, decreasing the number of packets in the queue by one, is the probability of a departure and no arrival or $s(1 - p)$. Finally, for the interior states, the

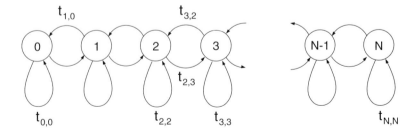

Fig. 3.6. Geom/Geom/1 queue state transition diagram

probability of staying in the same state between two slots is the sum of the probabilities of two events. One term is the probability of no arrival and no service completion or $(1 - p)(1 - s)$, and the other term is the probability of an arrival and a service completion in the same slot or ps.

The boundary states require some thought. Consider first state 0 and transition probability $t_{0,0}$. Then if one allows a packet to enter and leave an empty queue in the same slot [virtual cut through (Kermani 79)] $t_{0,0} = (1 - p) + ps$. This is because either no arrival or an arrival and immediate departure will leave the queue in the same state 0. Also $t_{0,1}$ is $p(1 - s)$ as stated above.

Now one could also implement the more traditional store-and-forward arrival policy where an arriving packet to an empty queue must wait one slot before departing. Thus $t_{0,0}$ is $(1 - p)$ and $t_{0,1}$ is p. That is, a packet arriving to an empty queue waits at least one slot prior to departure as the system always goes from state 0 to state 1 upon an arrival.

For the right boundary, let's suppose now that we have a finite buffer queue (N packets at most). In a single slot either departures precede arrivals or arrivals precede departures. The first case is preferable as a departing packet from a full (state N) queue will leave a space for a subsequent arrival in the same slot. In the latter case an arriving packet to a full queue may be turned away (lost or cleared) even though a space may become available later in the slot.

If departures preceed arrivals, $t_{N,N}$ is

$$t_{N,N} = p(s + (1 - s)) + (1 - p)(1 - s) \tag{3.33}$$

$$t_{N,N} = ps + p(1 - s) + (1 - p)(1 - s) \tag{3.34}$$

$$t_{N,N} = ps + (1 - s) \tag{3.35}$$

and

$$t_{N,N-1} = (1 - p)s \tag{3.36}$$

For $t_{N,N}$ the queue stays in the same state if either there is an arrival and a service completion or no service completion. The queue enters state $N - 1$ from state N if there is no arrival and a service completion.

It is important to realize that from state to state some transition must transpire, even if it results in no state change. The sum of all outgoing transition probabilities for a state must also sum to one. Thus

$$t_{N,N-1} + t_{N,N} = (1 - p)s + ps + (1 - s) = 1 \tag{3.37}$$

This is a good way to check whether one has the correct transition probabilities for any discrete time Markov chain. The bookkeeping can become burdensome though for larger chains.

Finally, suppose arrivals occur before departures in a slot. Then $t_{N,N-1} = s$ and $t_{N,N} = 1 - s$. That is, if there is no service departure (with probability $1 - s$), the queue stays in state N.

Summarizing for a queue with virtual cut through and departures before arrivals, one has

$$t_{0,0} = ps + (1 - p) \tag{3.38}$$

$$t_{0,1} = p(1 - s) \tag{3.39}$$

$$t_{n,n+1} = p(1 - s) \qquad 1 \le n \le N - 1 \tag{3.40}$$

$$t_{n,n-1} = s(1 - p) \qquad 1 \le n \le N \tag{3.41}$$

$$t_{n,n} = (1 - p)(1 - s) + ps \qquad 1 \le n \le N - 1 \tag{3.42}$$

$$t_{N,N} = ps + (1 - s) \tag{3.43}$$

To solve for the state probabilities, one can draw vertical boundaries, between adjacent states, and equate the flow of probability flux from left to right to that from right to left. For a discrete time queue, the probability flux flowing through a transition is the product of the transition probability of that transition and the state probability the transition originates from. This is not too different from the continuous time case.

One has

$$p(1 - s)P_0 = s(1 - p)P_1 \tag{3.44}$$

$$p(1 - s)P_1 = s(1 - p)P_2 \tag{3.45}$$

$$p(1 - s)P_2 = s(1 - p)P_3 \tag{3.46}$$

$$\cdot$$

$$\cdot$$

$$p(1 - s)P_{n-1} = s(1 - p)P_n \tag{3.47}$$

$$\cdot$$

$$\cdot$$

$$p(1 - s)P_{N-1} = s(1 - p)P_N \tag{3.48}$$

With algebra we can solve for the equilibrium state probabilities recursively (where, again, we use capital "P" for the equilibrium state probability to distinguish it from small "p," the arrival probability).

$$P_1 = \frac{p(1-s)}{s(1-p)} P_0 \tag{3.49}$$

$$P_2 = \frac{p(1-s)}{s(1-p)} P_1 \tag{3.50}$$

.

.

$$P_n = \frac{p(1-s)}{s(1-p)} P_{n-1} \tag{3.51}$$

.

.

$$P_N = \frac{p(1-s)}{s(1-p)} P_{N-1} \tag{3.52}$$

Chaining the equations together, one has

$$P_1 = \left(\frac{p(1-s)}{s(1-p)}\right) P_0 \tag{3.53}$$

$$P_2 = \left(\frac{p(1-s)}{s(1-p)}\right)^2 P_0 \tag{3.54}$$

$$P_n = \left(\frac{p(1-s)}{s(1-p)}\right)^n P_0 \tag{3.55}$$

$$P_N = \left(\frac{p(1-s)}{s(1-p)}\right)^N P_0 \tag{3.56}$$

Putting this together

$$P_n = \left(\frac{p(1-s)}{s(1-p)}\right)^n P_0 \qquad n = 1, 2 \ldots N \tag{3.57}$$

To solve for P_0 for an infinite buffer queue, one uses the normalization equation as with continuous time queues.

$$P_0 + P_1 + P_2 + \cdots + P_n + \cdots = 1 \tag{3.58}$$

Substituting

$$P_0 + \left(\frac{p(1-s)}{s(1-p)}\right) P_0 + \left(\frac{p(1-s)}{s(1-p)}\right)^2 P_0 + \cdots = 1 \tag{3.59}$$

or

$$P_0 \left(\sum_{i=0}^{\infty} \left(\frac{p(1-s)}{s(1-p)} \right)^i \right) = 1 \tag{3.60}$$

$$P_0 = 1 / \sum_{i=0}^{\infty} \left(\frac{p(1-s)}{s(1-p)} \right)^i \tag{3.61}$$

Using a summation formula from the appendix, one has for an infinite buffer Geom/Geom/1 queue with virtual cut through and departures occurring in a slot before arrivals in a slot

$$P_0 = 1 - \frac{p(1-s)}{s(1-p)} \tag{3.62}$$

For a finite buffer Geom/Geom/1/N queue with virtual cut through switching and departures in a slot occurring before arrivals, one does the same as above except the summation is finite. Using a different summation from the appendix

$$P_0 = 1 / \left(\sum_{i=0}^{N} \left(\frac{p(1-s)}{s(1-p)} \right)^i \right) \tag{3.63}$$

$$P_0 = \frac{1 - \frac{p(1-s)}{s(1-p)}}{1 - \left(\frac{p(1-s)}{s(1-p)} \right)^{N+1}} \tag{3.64}$$

For an infinite buffer system, one can substitute equation (3.62) into equation (3.57) to find the equilibrium state probabilities as a single function of p and s. In fact if

$$\rho = \frac{p(1-s)}{s(1-p)} \tag{3.65}$$

One then has

$$P_n = \rho^n (1 - \rho) \qquad n = 0, 1, 2 \ldots \tag{3.66}$$

This is the same form as the earlier expression for the equilibrium state probabilities of the M/M/1 queue [equation (3.31)]! Also, in much the same way as for an M/M/1 queue, the arrival probability p should be less than the service completion probability s for this infinite buffer discrete time queue's stability.

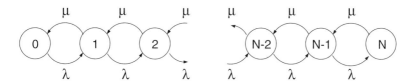

Fig. 3.7. Finite buffer M/M/1/N Markov chain

3.3 Some Important Single Queue Models

A number of useful single queue models and solutions are presented in this section.

3.3.1 The Finite Buffer M/M/1 Queueing System

Here we have an M/M/1 queue with the assumption that it holds at most N customers (one customer in the server, $N-1$ in the waiting line). If a customer arrives to a full queue it is turned away. Such rejected customers are sometimes called lost or cleared from the system. Note that this type of system is called an M/M/1/N queue where the fourth symbol is the buffer size.

The Markov chain of this model appears in Figure 3.7. We can draw vertical boundaries and equate the probability flux flowing from left to right to the flux flowing from right to left. Then with some simple algebra and solution chaining as in equations (3.11) through (3.18), one has

$$p_1 = \left(\frac{\lambda}{\mu}\right) p_0 \tag{3.67}$$

$$p_2 = \left(\frac{\lambda}{\mu}\right) p_1 = \left(\frac{\lambda}{\mu}\right)^2 p_0 \tag{3.68}$$

$$p_3 = \left(\frac{\lambda}{\mu}\right) p_2 = \left(\frac{\lambda}{\mu}\right)^3 p_0 \tag{3.69}$$

$$\vdots$$

$$p_n = \left(\frac{\lambda}{\mu}\right) p_{n-1} = \left(\frac{\lambda}{\mu}\right)^n p_0 \tag{3.70}$$

$$\vdots$$

$$p_N = \left(\frac{\lambda}{\mu}\right) p_{N-1} = \left(\frac{\lambda}{\mu}\right)^N p_0 \tag{3.71}$$

or

$$p_n = \left(\frac{\lambda}{\mu}\right)^n p_0 \qquad 1 \le n \le N \tag{3.72}$$

Using the normalization equation

$$p_0 + p_1 + p_2 + p_3 + \cdots + p_N = 1 \tag{3.73}$$

$$p_0 + \left(\frac{\lambda}{\mu}\right) p_0 + \left(\frac{\lambda}{\mu}\right)^2 p_0 + \left(\frac{\lambda}{\mu}\right)^3 p_0 + \cdots + \left(\frac{\lambda}{\mu}\right)^N p_0 = 1 \tag{3.74}$$

$$p_0 = \frac{1}{1 + \left(\frac{\lambda}{\mu}\right) + \left(\frac{\lambda}{\mu}\right)^2 + \left(\frac{\lambda}{\mu}\right)^3 + \cdots + \left(\frac{\lambda}{\mu}\right)^N} \tag{3.75}$$

$$p_0 = \frac{1}{\sum_{n=0}^{N} \left(\frac{\lambda}{\mu}\right)^n} \tag{3.76}$$

Using a summation from the appendix

$$p_0 = \frac{1 - \frac{\lambda}{\mu}}{1 - \left(\frac{\lambda}{\mu}\right)^{N+1}} \tag{3.77}$$

To obtain numerical values for a particular set of parameters, one uses the above formula to calculate p_0 and then substitutes p_0 into equation (3.72) for $p_1, p_2, p_3 \dots$.

This model is basically the same as that for the M/M/1 queue except for the number of states. Here there is a finite number of states, rather than the infinite number of states of the M/M/1 queue. The change in the number of states requires a renormalization that is allowed for in the denominator of equation (3.77). The numerator is the same as for the M/M/1 system. In fact, if the buffer size goes to infinity, this expression for p_0 reduces to $1 - \frac{\lambda}{\mu}$, the M/M/1 queue result. For a finite buffer M/M/1 queue, the arrival rate λ can be greater than the service rate μ since, if the queue fills up, customers are simply turned away.

3.3.2 The M/M/m/m Loss Queueing System

Suppose now, as Erlang did, that one has a bank of m parallel servers with negative exponential service times and no waiting line. (Figure 3.8). That is, a (Poisson) arriving customer is placed in an empty server. If all servers are busy, though, an arriving customer is cleared (lost) from the system. This

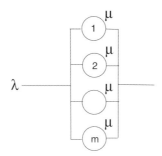

Fig. 3.8. M/M/m/m loss system schematic

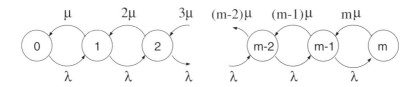

Fig. 3.9. M/M/m/m loss system Markov chain

model may represent m telephone circuits fed by a common pool of users. If a call doesn't immediately get through, it is "blocked." Naturally this model doesn't take into account retries of blocked calls.

The state transition diagram appears in Figure 3.9. Note that if there are n customers in the system, the aggregate transition rate is $n\mu$ (n busy servers complete service n times faster than one busy server). The solution technique for the equilibrium state probabilities is the same as usual. One draws vertical boundaries between states and equates the flow of probabilities flux across the boundaries

$$\lambda p_0 = \mu p_1 \tag{3.78}$$

$$\lambda p_1 = 2\mu p_2 \tag{3.79}$$

$$\lambda p_2 = 3\mu p_3 \tag{3.80}$$

$$.$$

$$.$$

$$\lambda p_n = (n+1)\mu p_{n+1} \tag{3.81}$$

$$.$$

$$\lambda p_{m-1} = m\mu p_m \tag{3.82}$$

With some algebra and solution chaining, one has, where $n \leq m$

$$p_1 = \left(\frac{\lambda}{\mu}\right) p_0 \tag{3.83}$$

$$p_2 = \left(\frac{\lambda}{2\mu}\right) p_1 = \frac{1}{2} \left(\frac{\lambda}{\mu}\right)^2 p_0 \tag{3.84}$$

$$p_3 = \left(\frac{\lambda}{3\mu}\right) p_2 = \frac{1}{6} \left(\frac{\lambda}{\mu}\right)^3 p_0 \tag{3.85}$$

.

.

.

$$p_n = \left(\frac{\lambda}{n\mu}\right) p_{n-1} = \frac{1}{n!} \left(\frac{\lambda}{\mu}\right)^n p_0 \tag{3.86}$$

.

.

$$p_m = \left(\frac{\lambda}{m\mu}\right) p_{N-1} = \frac{1}{m!} \left(\frac{\lambda}{\mu}\right)^m p_0 \tag{3.87}$$

Using the normalization equation

$$p_0 + p_1 + p_2 + p_3 + \cdots + p_N = 1 \tag{3.88}$$

$$p_0 + \left(\frac{\lambda}{\mu}\right) p_0 + \frac{1}{2} \left(\frac{\lambda}{\mu}\right)^2 p_0 + \frac{1}{6} \left(\frac{\lambda}{\mu}\right)^3 p_0 + \cdots + \frac{1}{m!} \left(\frac{\lambda}{\mu}\right)^m p_0 = 1 \tag{3.89}$$

$$p_0 = \frac{1}{1 + \left(\frac{\lambda}{\mu}\right) + \frac{1}{2} \left(\frac{\lambda}{\mu}\right)^2 + \frac{1}{6} \left(\frac{\lambda}{\mu}\right)^3 + \cdots + \frac{1}{m!} \left(\frac{\lambda}{\mu}\right)^m} \tag{3.90}$$

$$p_0 = \frac{1}{1 + \sum_{n=1}^{m} \frac{1}{n!} \left(\frac{\lambda}{\mu}\right)^n} \tag{3.91}$$

In telephone system design an important performance measure is the probability that all servers are busy. This is the fraction of time that arriving customers are turned away. Naturally this performance measure is simply p_m. So

$$p_m = \frac{\frac{1}{m!}\left(\frac{\lambda}{\mu}\right)^m}{1 + \sum_{n=1}^{m} \frac{1}{n!}\left(\frac{\lambda}{\mu}\right)^n} \tag{3.92}$$

This formula is known as the Erlang B formula or Erlang's formula of the first kind in Europe.

A related queueing model arises if every arriving customer gets a server (i.e., there is an unlimited number of servers). For this M/M/1/∞ system, it is still true that

$$p_n = \frac{1}{n!}\left(\frac{\lambda}{\mu}\right)^n p_0 \tag{3.93}$$

Also from the equation above for p_0, letting $m \to \infty$

$$p_0 = \frac{1}{1 + \sum_{n=1}^{\infty} \frac{1}{n!}\left(\frac{\lambda}{\mu}\right)^n} = e^{-\frac{\lambda}{\mu}} \tag{3.94}$$

Here we have used a summation from the appendix.

3.3.3 M/M/m Queueing System

Now suppose we have the model of the previous section with m servers and with a waiting line (Figure 3.10). This is a good model of a telephone system where one queues to get an outside line. The Markov chain appears in Figure 3.11. Note that the chain transitions are identical to those of the M/M/1/M model below state m. Above state m all departure transitions have value $m\mu$. This is because if there are m or more customers in the system at most m servers are busy, which leads to an aggregate departure rate of $m\mu$. The buffer here is unlimited (infinite) in size.

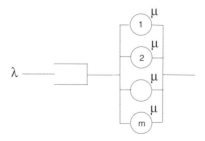

Fig. 3.10. M/M/m queueing system schematic

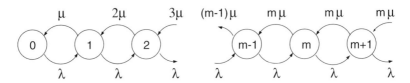

Fig. 3.11. M/M/m queueing system Markov chain

For the state probabilities at or below state m $(n \leq m)$, one has from the M/M/m/m system

$$p_n = \frac{1}{n!} \left(\frac{\lambda}{\mu} \right)^n p_0 \qquad n \leq m \qquad (3.95)$$

For states above state m

$$\lambda p_m = m\mu p_{m+1} \qquad (3.96)$$

$$\lambda p_{m+1} = m\mu p_{m+2} \qquad (3.97)$$

$$\lambda p_{m+2} = m\mu p_{m+3} \qquad (3.98)$$

$$.$$

$$.$$

or

$$p_{m+1} = \frac{1}{m} \left(\frac{\lambda}{\mu} \right) p_m \qquad (3.99)$$

$$p_{m+2} = \frac{1}{m} \left(\frac{\lambda}{\mu} \right) p_{m+1} = \frac{1}{m^2} \left(\frac{\lambda}{\mu} \right)^2 p_m \qquad (3.100)$$

$$p_{m+3} = \frac{1}{m} \left(\frac{\lambda}{\mu} \right) p_{m+2} = \frac{1}{m^3} \left(\frac{\lambda}{\mu} \right)^3 p_m \qquad (3.101)$$

$$.$$

$$.$$

or

$$p_n = \left(\frac{\lambda}{m\mu} \right)^{n-m} p_m \qquad n > m \qquad (3.102)$$

So

$$p_n = \left(\frac{1}{m!} \left(\frac{\lambda}{\mu} \right)^m \right) \left(\left(\frac{\lambda}{m\mu} \right)^{n-m} \right) p_0 \qquad n > m \qquad (3.103)$$

$$p_n = \frac{1}{m!m^{n-m}} \left(\frac{\lambda}{\mu} \right)^n p_0 \qquad n > m \qquad (3.104)$$

One can then substitute the two equations for p_n ($n \leq m$, $n > m$) into the normalization equation

$$p_0 + p_1 + p_2 + \cdots + p_n + \cdots = 1 \qquad (3.105)$$

Solving for p_0 one obtains after some algebra and using a summation from the appendix

$$p_0 = \left[1 + \sum_{n=1}^{m-1} \frac{1}{n!} \left(\frac{\lambda}{\mu} \right)^n + \frac{1}{m!} \left(\frac{\lambda}{\mu} \right)^m \left(\frac{1}{1-\rho} \right) \right]^{-1} \qquad (3.106)$$

Here $\rho = \lambda/m\mu$. In telephone system design, another important performance measure, for this M/M/m model, is the probability that a call doesn't get a server immediately but must wait. This is equal to the sum of the probabilities that there are m, $m+1$, $m+2 \ldots$ customers.

$$\text{Prob[queueing]} = \sum_{n=m}^{\infty} p_n \qquad (3.107)$$

Since only states $n > m$ are involved, the resulting Erlang C (or formula of the second kind in Europe) formula can be found as

$$\text{Prob[queueing]} = \left(\sum_{n=m}^{\infty} \frac{1}{m!m^{n-m}} \left(\frac{\lambda}{\mu} \right)^n \right) p_0 \qquad (3.108)$$

$$\text{Prob[queueing]} = \frac{1}{m!} \left(\frac{\lambda}{\mu} \right)^m \left(\frac{1}{1-\rho} \right) p_0 \qquad (3.109)$$

or

$$\text{Prob[queueing]} = \frac{\frac{1}{m!} \left(\frac{\lambda}{\mu} \right)^m \left(\frac{1}{1-\rho} \right)}{\left[1 + \sum_{n=1}^{m-1} \frac{1}{n!} \left(\frac{\lambda}{\mu} \right)^n + \frac{1}{m!} \left(\frac{\lambda}{\mu} \right)^m \left(\frac{1}{1-\rho} \right) \right]} \qquad (3.110)$$

Again, $\rho = \lambda/m\mu$.

3.3.4 A Queueing-Based Memory Model

The previous queueing models are well suited to model customers (e.g., jobs and packets) waiting in a line for some service. A clever model from Kaufman (Kaufman 81, Ross 97) can also take a form of memory requirements into account. In this model the aggregate arrival stream is Poisson with mean arrival rate λ. An arriving customer belongs to the ith of k customer classes with independent probability q_i. A customer of the ith class has a distinct temporal (service time) requirement τ_i and memory space requirement b_i. There are C units of memory. An arriving customer with a memory requirement of b_i units is accommodated if there are at least b_i (not necessarily contiguous) units of free memory; otherwise it is blocked and cleared from the system. The residency time distribution of a class can have a rational Laplace transform with mean $1/\mu_i$. Thus the time distribution belongs to a class that includes, but is more general than, a simple negative exponential distribution.

The genius of Kaufman's model lies in the simplicity of the solution for the equilibrium state probabilities. Following Kaufman's notation, let

$$\underline{n}_i = (n_1, n_2, \ldots n_{i-1}, n_i, n_{i+1}, \ldots n_k) \tag{3.111}$$

This is the (population) vector of the number of customers of each of the K classes in the queueing system. Then let

$$\underline{n}_i^+ = (n_1, n_2, \ldots n_{i-1}, n_i + 1, n_{i+1}, \ldots n_k) \tag{3.112}$$

$$\underline{n}_i^- = (n_1, n_2, \ldots n_{i-1}, n_i - 1, n_{i+1}, \ldots n_k) \tag{3.113}$$

Here $\underline{n}_i^+$ has one additional customer in the ith queue compared with $\underline{n}_i$ and $\underline{n}_i^-$ has one less customer in the ith queue compared with $\underline{n}_i$.

Let Ω be the set of allowable states that depends on the resource sharing policy being used. Also

$$\delta_i^+(\underline{n}) = \begin{cases} 1 \text{ if } \underline{n}_i^+ \in \Omega \\ 0 \text{ otherwise} \end{cases} \tag{3.114}$$

$$\delta_i^-(\underline{n}) = \begin{cases} 1 \text{ if } \underline{n}_i^- \in \Omega \\ 0 \text{ otherwise} \end{cases} \tag{3.115}$$

Here $\underline{n}_i^+ \in \Omega$ means that the state $\underline{n}_i^+$ is a member of Ω. Naturally

$$\underline{n} \cdot \underline{b} = \sum_{i=1}^{k} n_i b_i \tag{3.116}$$

Here $\underline{n} \cdot \underline{b}$ is the amount of memory space that is occupied in state $\underline{n}$.

The Markov chain state transition diagram is more complex than in the previous cases. To solve for the equilibrium state probabilities, one can start by writing a global balance equation for each state $\underline{n}$ belonging to set Ω.

$$\left[\sum_{i=1}^{k} \lambda_i \delta_i^+(\underline{n}) + \sum_{i=1}^{k} n_i \mu_i \delta_i^-(\underline{n}) \right] p(\underline{n})$$

$$= \sum_{i=1}^{k} \lambda_i \delta_i^-(\underline{n}) p(\underline{n}_i^-) + \sum_{i=1}^{k} (n_i + 1) \mu_i \delta_i^+(\underline{n}) p(\underline{n}_i^+) \qquad (3.117)$$

Here $\lambda_i = \lambda q_i$. The left side of this equation is associated with the probability flux leaving state $\underline{n}$. The right side of this equation is associated with the probability flux entering state $\underline{n}$. The δ terms account for state space boundaries. At an interior state all δ's have a value of 1.0. More importantly in terms of a solution for the equilibrium state probabilities, one can write a set of local balance equations for each state $\underline{n}$ belonging to Ω

$$\lambda_i \delta_i^-(\underline{n}) p(\underline{n}_i^-) = n_i \mu_i \delta_i^-(\underline{n}) p(\underline{n}) \qquad i = 1, 2 \ldots k \qquad (3.118)$$

The left side of this equation represents the probability flux from an arriving class i customer causing the system to enter state $\underline{n}$. The right side represents a class i customer departing the system and causing the system to leave state $\underline{n}$. Again, the δ's account for state space boundaries. These local balance equations can be solved recursively

$$p(\underline{n}) = \frac{\lambda_i}{n_i \mu_i} p(n_1, \ldots n_{i-1}, n_i - 1, n_{i+1}, \ldots n_k) \qquad (3.119)$$

Proceeding with the recursion to zero out the ith term

$$p(\underline{n}) = \frac{a_i^{n_i}}{n_i!} p(n_1, \ldots n_{i-1}, 0, n_{i+1}, \ldots n_k) \qquad (3.120)$$

Here $a_i = \lambda_i / \mu_i$. If one does this for each class, one arrives at the equilibrium state probability solution

$$p(\underline{n}) = \left(\prod_{i=1}^{k} \frac{a_i^{n_i}}{n_i!} \right) p(0, 0, 0 \ldots 0)$$

$$p(\underline{n}) = \left(\prod_{i=1}^{k} \frac{a_i^{n_i}}{n_i!} \right) G^{-1}(\Omega) \qquad all \ \underline{n} \in \Omega \qquad (3.121)$$

In the above equation, $G^{-1}(\Omega)$ is the inverse normalization constant from the normalization equation

$$G(\Omega) = \sum_{\underline{n} \in \Omega} \left(\prod_{i=1}^{k} \frac{a_i^{n_i}}{n_i!} \right) \qquad (3.122)$$

Normalization constants are correction factors used so that the sum of a finite number of state probabilities does indeed sum to one. The solution above can be inserted into either the local or global balance equation and the balancing can be verified to prove we have the correct solution. One can compute the probability that a class i arriving customer is blocked from

$$p_{b_i} = \sum_{\underline{n} \in B_i^+} p(\underline{n}) \tag{3.123}$$

Here

$$B_i^+ = \{\underline{n} \in \Omega : \quad \underline{n}_i^+ \quad \text{not in} \quad \Omega\} \tag{3.124}$$

Here B_i^+ is the set of states where the system is in a blocking state.

3.3.5 M/G/1 Queueing System

The previous continuous queueing system had Poisson arrivals and negative exponentially distributed service times. With these assumptions, one has a memoryless system and the solution for the equilibrium state probabilities is straightforward. With a memoryless system the state of the queueing system at any point in time is simply the number of customers in the queueing system at the time. This makes calculation tractable. One need not account for the time since the last arrival or the time the customer in the server has been in service so far.

Assuming non-exponential style arrival or departure times does indeed make analysis more complex. However there are some special cases where relatively simple results are available. One such case is the M/G/1 queue. Here we have a single queue where there are Poisson arrivals and a "general" (arbitrary) service time distribution. Even if we assume any possible (a general) service time distribution, as the M/G/1 queue does, it is possible to develop a fairly simple formula for the average (expected) number of customers in the queueing system. This is the Pollaczek and Khinchin mean value formula published by these researchers in 1930 (Pollaczek) and 1932 (Khinchin), respectively. Although they arrived at this result by complex means, David Kendall in 1951 published a simple deviation that we will outline (Gross 85, Hammond 86, Kleinrock 75, Robertazzi 00).

Kendall's Approach and Result

Kendall's approach is to use a Markov chain "imbedded" at the departure instants. This is based on the idea that, for some queueing systems, the queue behavior in equilibrium at an arbitrary point in time t is the same as the behavior at the departure points (Papoulis 02). A proof of this can first show that the equilibrium state probabilities "seen" at the departure instants are identical to the equilibrium state probabilities seen at the arrival instants. However since arrivals are Poisson (and occur at random times that are independent of the queueing system state) one may go further and show that the equilibrium state probabilities at the arrival instants are identical to those at any point in time.

Kendall's approach is to write a recursion for the number of customers at the departure instants. If the queue is nonempty at the ith departure instant ($n_i > 0$)

$$n_{i+1} = n_i - 1 + a_{i+1} \qquad n_i > 0 \qquad (3.125)$$

Here n_i is the number of customers in the queue immediately after the ith departure instant. The "-1" accounts for the departure at the $(i + 1)$st instant. Also a_{i+1} is the number of customers that arrive into the system between the ith and $(i + 1)$st departure instants. Naturally if the queue is empty after the ith departure instant, one has

$$n_{i+1} = a_{i+1} \qquad n_i = 0 \qquad (3.126)$$

We'd like to combine these two equations into a single equation. This can be done with a unit step function

$$u(n_i) = \begin{cases} 1 & n_i > 0 \\ 0 & n_i = 0 \end{cases} \qquad (3.127)$$

So

$$\boxed{n_{i+1} = n_i - u(n_i) + a_{i+1}} \qquad (3.128)$$

Kendall's approach is to square both sides of the recursion and then to take the expectations of both sides

$$E[(n_{i+1})^2] = E[(n_i - u(n_i) + a_{i+1})^2] \qquad (3.129)$$

Expanding the right side, one has

$$E[(n_{i+1})^2] = E[n_i^2] + E[(u(n_i))^2] + E[a_{i+1}^2]$$

$$-2E[n_i u(n_i)] + 2E[n_i a_{i+1}] - 2E[u(n_i)a_{i+1}] \qquad (3.130)$$

If one solves for each of these terms (Gross 85, Kleinrock 75, Robertazzi 00), one arrives with some algebra at

$$E[n] = \frac{2\rho - \rho^2 + \lambda^2 \sigma_s^2}{2(1 - \rho)} \qquad (3.131)$$

$$E[n] = \rho + \frac{\rho^2 + \lambda^2 \sigma_s^2}{2(1 - \rho)} \qquad (3.132)$$

These are two forms of the Pollaczek–Khinchin (P–K) mean value formula. All one needs is the arrival rate λ, the utilization ρ (mean arrival rate divided by mean service rate of the general distribution $b(t)$ or $\rho = 1 - p_0$), and the variance of the service distribution used. Thus, for any distribution of service time, only its first two moments are needed to evaluate the mean number of customers in the queue. This is surprising as an arbitrary service time distribution needs higher moments to completely specify it.

Naturally the mean delay a customer experiences in passing through the queue can be calculated from the P–K formula and Little's Law.

The M/G/1 State Transition Diagram

An interesting question is finding the topology of the Markov chain imbedded at departure instants. We will do this by first creating a matrix of state transition probabilities. Let

$$\underline{P} = [P_{rs}] = P[n_{i+1} = s | n_i = r] \qquad (3.133)$$

To move from r customers in the queueing system immediately after the ith departure instant ($n_i = r$) to s customers immediately after the $(i+1)$st departure instant, there should be $s - r + 1$ arrivals between the two departure instants. We need the "+1" terms as the queue loses one customer at the $(i+1)$st departure instant. Thus

$$[P_{rs}] = k_{s-r+1} = k_{\#\text{arrivals}} \qquad (3.134)$$

Continuing with Kendall's notation, one has the following (infinite size) matrix

$P = [P_{rs}]$	0	1	2	3	4	.
0	k_0	k_1	k_2	k_3	k_4	.
1	k_0	k_1	k_2	k_3	k_4	.
2	0	k_0	k_1	k_2	k_3	.
3	0	0	k_0	k_1	k_2	.
4	0	0	0	k_0	k_1	.
.	.	.	.	.	.	.

This infinite size matrix is a stochastic matrix. That is, the sum of entries in any row is one. This is because the probability of going from a specific state (row number) to *some* state (column number) is 1.0.

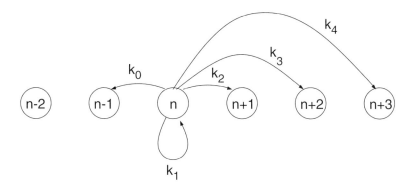

Fig. 3.12. M/G/1 (embedded Markov chain) state transition diagram

To see how the entries were placed in the table, consider entry $[p_{2,4}]$ (row 2, column 4). To go from two customers immediately after a departure instant to having four customers immediately after the next departure instant requires three arrivals (or k_3) as one customer departs after the $(i + 1)$st departure instant.

A partial Markov chain (state transition diagram) is shown in Figure 3.12. Here transitions leaving state n are illustrated.

How does one compute the k_j's? One has

$$k_j = \int_0^\infty \text{Prob}[j \text{ arrivals} \mid \text{time } s]b(s)ds \qquad (3.135)$$

Here we integrate over the probability that there are j arrivals in a time s between consecutive departures where this probability is weighted by the distribution of s. That is, $b(s)$ is the general service time distribution. Since arrivals are Poisson

$$k_j = \int_0^\infty \frac{(\lambda s)^j e^{-\lambda s}}{j!} b(s)ds \qquad (3.136)$$

There are two ways to solve for the equilibrium state probabilities of this M/G/1 embedded Markov chain. One is by drawing vertical boundaries between adjacent states, equating flow in both directions across the boundaries, and creating a (somewhat complex) recursive series of equations starting from state 0 (Robertazzi 00). In theory one can solve for all state probabilities as a function of p_0 and then normalize the probabilities.

The problem with this numerical approach is that it is approximate in that only a finite number of transitions can be used leaving each state, on a computer, not the infinite number of states for which the mathematical model calls. An alternative approach to solve for the chain's state probabilities is to use moment generating functions. Moment generating functions are a frequency spectrum-like description of probability distributions.

Advanced books on queueing theory (Gross 85, Robertazzi 00, Kleinrock 75) show that, if K(z) is the moment generating function of the k_j

$$K(z) = \sum_{j=0}^{\infty} k_j z^j \qquad (3.137)$$

Then

$$\Pi(z) = \frac{(1-\rho)(1-z)K(z)}{K(z) - z} \qquad (3.138)$$

Here $\Pi(z)$ is the moment generating function of the chain equilibrium state probabilities. In general $\Pi(z)$ can be inverted (analytically or sometimes only numerically) to find the equilibrium state probabilities.

3.4 Common Performance Measures

A great deal of effort has been expended in the previous pages of this chapter in determining the equilibrium state probabilities. The main reason we are interested in these probabilities is that many common performance measures are simple functions of these state probabilities. Although this is often not the only or the most efficient means of computing these performance measures, it is the most direct way. In this subsection, continuous time queue performance measures will be described. Analogous expressions hold for discrete time queues. For instance, for a queue with a single server, the fraction of time that the server is busy, or the utilization of the server, is

$$U = 1 - p_0 \qquad (3.139)$$

If a single queue holds at most N customers, the probability an arriving customer is "blocked" from entering the queue because it is full, or the blocking probability, is

$$P_B = p_N \qquad (3.140)$$

The mean (average) number of customers in an infinite size buffer $\bar{n}$ is

$$\bar{n} = \sum_{n=1}^{\infty} n p_n \qquad (3.141)$$

This is a weighted average of each possible number of customers and the probability in equilibrium that there are that number of customers in the queue. Note that it would make no difference if the index of the summation started at $n = 0$ since the zeroth term would have a value of zero.

The mean "throughput" or flow of customers through a single-server, infinite buffer size queue is

$$\overline{T} = \sum_{n=1}^{\infty} \mu(n)p_n \tag{3.142}$$

In this equation $\mu(n)$ is the state-dependent service rate of the server when there are n customers. Again, the throughput expression is a weighted sum of the (state-dependent) service rate when there are n customers multiplied by the equilibrium probability that there are n customers. Since the throughput of an empty queue is zero, the summation index starts at $n = 1$. If one has a finite buffer queue of size N, then the equations for $\overline{n}$ and $\overline{T}$ are the same except that ∞ is replaced by N.

Finally, consider an infinite size buffer queue so that the mean arrival rate equals the mean throughput. Then from Little's Law, the mean delay $\overline{\tau}$ a customer experiences in moving through the queue (and server) is

$$\overline{\tau} = \frac{\overline{n}}{\lambda} = \frac{\overline{n}}{\overline{T}} = \frac{\sum_{n=1}^{\infty} np_n}{\sum_{n=1}^{\infty} \mu(n)p_n} \tag{3.143}$$

Note that the units of this ratio are [customers] divided by [customers/second] or [seconds], which makes sense for mean delay.

3.5 Markovian Queueing Networks

The queueing modeled in the previous sections involves a single queue. What about networks of queues? Networks of queues can be either be open (with external arrivals and departures from the network) or closed (sealed) with a fixed number of customers circulating in the network. In fact, for both types of continuous time queueing networks, the elegant solutions of the previous sections for equilibrium state probabilities can be generalized into what is referred to as the product form solution.

For a continuous time network of M Markovian queues, we seek an expression for $p(\underline{n})$ the equilibrium probability that the network is in state $\underline{n}$. Here $\underline{n}$ is a vector

$$\underline{n} = (n_1, n_2, \ldots, n_{i-1}, n_i, n_{i+1}, \ldots, n_M) \tag{3.144}$$

In this vector n_i is the number of customers in the ith queue in state $\underline{n}$. We assume a Markovian system (i.e., Poisson arrivals for open networks and independent negative exponential random service times for both open and closed networks). For Markovian systems, the system state at any time instant is completely summarized by the number of customers in each network queue at that time instant. Again, the system is memoryless so that one does not have to include the times customers have been in service, or the times since the last arrivals, into the system state.

Because even moderately sized networks of queues have very large numbers of states, the use of global balance equations to solve for the equilibrium state probabilities would be impractical as one would have to solve sets of linear equations that are too large. However, the existence of tractable product form solutions for Markovian queueing network equilibrium probabilities is in fact closely related to the concept of local balance. Recall that global balance equates the total flow of probability flux into a state on all incoming transitions to the total flow out of the state on all outgoing transitions. Local balance, on the other hand, equates the incoming and outgoing flows in subsets of a state's transitions.

Local balance exists only for certain classes of queueing networks, including open and closed Markovian networks where the queues have unlimited buffers and routing between queues is random. These are the two network models addressed in this section. It turns out that, if local balance exists, so does a product form solution and the reverse is true as well.

In terms of the history of product form results, Jackson in 1957 was the first to solve the open network problem. Ten years later, in 1967, Gordon and Newell provided a solution of the closed network problem. A classic paper generalizing these results is by Baskett, Chandy, Muntz, and Palacios in 1975.

The following two subsections discuss open and closed Markovian queueing networks in turn. We use a similar method to Kobayashi 78 and Schwartz 87.

3.5.1 Open Networks

Consider the open queueing network in Figure 3.13. A Markovian system of M queues, an external source of Poisson arriving customers, and an external destination for customers is pictured. Buffers are unlimited in size.

Customers are generated at the external source with mean rate λ. The probability that a customer generated at the external source is randomly

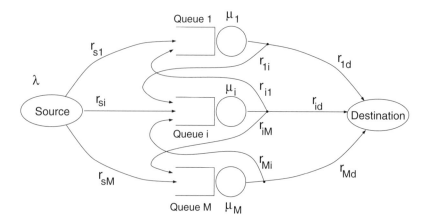

Fig. 3.13. Open queueing network schematic representation

chosen to enter queue i is r_{si}. Also, the (negative exponential distribution based) service rate of the ith queue is μ_i. The probability that a customer departing queue i enters queue j is r_{ij}. Customers may reenter the same queue they depart from (with probability r_{ii} for the ith queue). The probability that a customer leaving the ith queue proceeds to the external destination is r_{id}. In addition to the state vector [see equation (3.144) above], let

$$\underline{1}_i = (0, 0, \ldots, 0, 1, 0 \ldots, 0) \tag{3.145}$$

Here there is a "1" in the ith position.

Using this notation one can write a global balance equation for state $\underline{n}$ in an open network as

$$\left(\lambda + \sum_{i=1}^{M} \mu_i \right) p(\underline{n}) = \sum_{i=1}^{M} \lambda r_{si} p(\underline{n} - \underline{1}_i) \tag{3.146}$$

$$+ \sum_{i=1}^{M} \mu_i r_{id} p(\underline{n} + \underline{1}_i) + \sum_{i=1}^{M} \sum_{j=1}^{M} \mu_j r_{ji} p(\underline{n} + \underline{1}_j - \underline{1}_i)$$

This global balance equation equates the net flow of probability flux out of the state (left side) to the net flow into the state (right side). More specifically, the left side of the equation is associated with customers entering or leaving queues, causing the network to leave state $\underline{n}$. The right-hand side of the equation has three terms associated with the three ways that the network may directly enter state $\underline{n}$. One is via an external arrival to the ith queue when the network is in state $\underline{n} - \underline{1}_i$ (almost state $\underline{n}$ except that there are $n_i - 1$ customers in the ith queue). The second is via a departure from queue i to the destination when the network is in state $\underline{n} + \underline{1}_i$ (almost state $\underline{n}$ except that the ith queue has $n_i + 1$ customers). The last terms corresponds to a transfer between the jth and ith queues starting from state $\underline{n} + \underline{1}_j - \underline{1}_i$ and through a single transfer ending in state $\underline{n}$.

To obtain a local balance equation leading to a solution for the equilibrium probabilities, let's first consider what are called the traffic equations of the network—a simple way to calculate queue mean throughput in open networks. Let θ_i be the mean throughput of queue i. Then

$$\theta_i = r_{si} \lambda + \sum_{j=1}^{M} r_{ji} \theta_j \qquad i = 1, 2 \ldots M \tag{3.147}$$

This equation states that the ith queue mean throughput equals the sum of the mean external arrival rate to the ith queue and the mean rates of customer transfers from other queues to the ith queue. The M traffic equations are linear equations that can be solved by standard means for the M queues' throughputs. As an example, consider Figure 3.14.

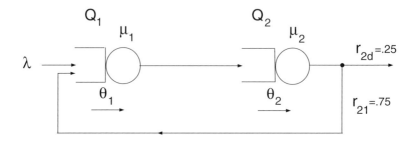

Fig. 3.14. Open queueing network example

The traffic equations of this network are

$$\theta_1 = \lambda + \frac{3}{4}\theta_2 \tag{3.148}$$

$$\theta_2 = \theta_1 \tag{3.149}$$

So solving

$$\theta_1 = \lambda + \frac{3}{4}\theta_1 \tag{3.150}$$

$$\theta_1 = \theta_2 = 4\lambda \tag{3.151}$$

Note that, since there is a feedback, that is, a customer may visit a queue several times, queue throughput can be larger than the arrival rate.

Now, to generate a local balance equation, the traffic equation may be solved for λr_{si}, which is then substituted into the global balance equation (3.146). After some algebraic manipulation (see Kobayashi 78, Robertazzi 00) one has the following local balance equation

$$\theta_i p(\underline{n} - \underline{1}_i) = \mu_i p(\underline{n}) \tag{3.152}$$

This equation states that for queue i the flow out of state $\underline{n}$ from a departure (right side) equals the flow from an arrival to queue i (left side) when the network is in state $\underline{n} - \underline{1}_i$ (state $\underline{n}$ with one less customer in queue i). Here the queues' total arrival rates are equal to the throughputs', the θ_i's. This balancing is not at all an obvious fact and is part of the amazing nature of local balance. It should be noted that local balance occurs in certain queueing networks (and generally not in electric circuits) because state transitions (unlike impedances) are labeled in a patterned manner that makes local balance possible. The equation can be rewritten as

$$p(\underline{n}) = \frac{\theta_i}{\mu_i} p(\underline{n} - \underline{1}_i) \tag{3.153}$$

Expanding this

$$p(\underline{n}) = \frac{\theta_i}{\mu_i} p(n_1, n_2, \ldots, n_{i-1}, n_i - 1, n_{i+1}, \ldots, n_M) \tag{3.154}$$

Repetitively using the local balance equation results in

$$p(\underline{n}) = \left(\frac{\theta_i}{\mu_i}\right)^{n_i} p(n_1, n_2, \ldots, n_{i-1}, 0, n_{i+1}, \ldots, n_M) \tag{3.155}$$

Here the ith term has been zeroed out. This can be done for each term resulting in our first version of the product form solution.

$$p(\underline{n}) = \left(\prod_{i=1}^{M} \left(\frac{\theta_i}{\mu_i}\right)^{n_i}\right) p(0, 0, 0, 0, 0) \tag{3.156}$$

$$p(\underline{n}) = \left(\prod_{i=1}^{M} \left(\frac{\theta_i}{\mu_i}\right)^{n_i}\right) p(\underline{0}) \tag{3.157}$$

To solve for $p(\underline{0})$, one can use the normalization equation

$$\sum_{\underline{n}} p(\underline{n}) = \sum_{\underline{n}} \left(\prod_{i=1}^{M} \left(\frac{\theta_i}{\mu_i}\right)^{n_i}\right) p(\underline{0}) \tag{3.158}$$

The index of the product term doesn't depend on $\underline{n}$ so that the summation and product can be interchanged.

$$\sum_{\underline{n}} p(\underline{n}) = \prod_{i=1}^{M} \left(\sum_{n_i=0}^{\infty} \left(\frac{\theta_i}{\mu_i}\right)^{n_i}\right) p(\underline{0}) \tag{3.159}$$

Using a summation formula from the appendix

$$\sum_{\underline{n}} p(\underline{n}) = \prod_{i=1}^{M} \left(1 - \frac{\theta_i}{\mu_i}\right)^{-1} p(\underline{0}) \tag{3.160}$$

Since $\sum_{\underline{n}} p(\underline{n}) = 1$

$$p(\underline{0}) = \prod_{i=1}^{M} \left(1 - \frac{\theta_i}{\mu_i}\right) \tag{3.161}$$

Writing this out in a final form

$$p(\underline{n}) = \prod_{i=1}^{M} p_i(n_i) \tag{3.162}$$

$$p_i(n_i) = \left(1 - \frac{\theta_i}{\mu_i}\right)\left(\frac{\theta_i}{\mu_i}\right)^{n_i} \tag{3.163}$$

One can see that the name "product form" comes from the fact the expression for the equilibrium probability of state $\underline{n}$ for an open Markovian network of queues with unlimited buffers and random routing is a product of terms. Each term is analogous to the expression for the state probabilities of an M/M/1 queue [equation (3.31)] except that λ is replaced by θ_i. The product form solution for the previous example is thus

$$p(n_1, n_2) = \left(1 - \frac{4\lambda}{\mu_1}\right)\left(\frac{4\lambda}{\mu_1}\right)^{n_1}\left(1 - \frac{4\lambda}{\mu_2}\right)\left(\frac{4\lambda}{\mu_2}\right)^{n_2} \tag{3.164}$$

Note that, if the buffers were finite and the routing was not random (e.g., a join the shortest queue policy for a bank of parallel queues), then there would be no local balance and no product form solution.

One can compute numerical probabilities using the product form equations (3.162) and (3.163), although one must first solve the traffic equations for the θ_i's. It is interesting to note that the form of equation (3.162) above is similar to a joint distribution of independent random variables. Does this mean the number of customers in the queues in an open Markovian network are independent of one another? This is only true at one instant at time. As the queueing theorist R. Disney (Disney 87) points out, if one compared the system state at two close instants, one would see significant correlations between the number of customers in different queues between the two instants.

3.5.2 Closed Networks

Consider now a closed Markovian queueing network. It is a "sealed" system with a fixed number of customers N circulating though it. The service rate of the ith queue is μ_i. Routing between queues is again random (r_{ij} is the probability that a customer departing the ith queue enters the jth queue). Buffers are big enough to always be able to accommodate all customers. A global balance equation for state $\underline{n}$ can be written as

$$\sum_{i=1}^{M} \mu_i p(\underline{n}) = \sum_{i=1}^{M}\sum_{j=1}^{M} \mu_j r_{ji} p(\underline{n} + \underline{1}_j - \underline{1}_i) \tag{3.165}$$

The formula equates the total probability flux leaving state $\underline{n}$ (left side) to that entering the state (right side). More specifically the left side is associated with departures from queues causing the network to leave state $\underline{n}$. The right side is associated with transfers from the jth to the ith queue when the network is in state $\underline{n} + \underline{1}_j - \underline{1}_i$ (essentially state $\underline{n}$ with one extra customer in queue j and one less customer in queue i) causing it to transit to network state $\underline{n}$.

Once again traffic equations can be written as

$$\theta_i = \sum_{j=1}^{M} r_{ji}\theta_j \qquad i = 1, 2 \ldots M \qquad (3.166)$$

Each of the above M equations equates the mean throughput of queue i, θ_i, to the sum of the mean customer flows from each of the queues entering queue i. Note that there is no external arrival term as there is for the traffic equations of open networks. Also, these traffic solutions do not have a unique solution. If $(\theta_1, \theta_2 \ldots, \theta_M)$ is a solution, so is $(c\theta_1, c\theta_2 \ldots, c\theta_M)$. What these equations yield are relative, not absolute, mean queue throughputs. To obtain a local balance equation, one combines the traffic equation with the global balance equation (Robertazzi 00). With some algebra (Robertazzi 00), the following local balance equation is obtained

$$\mu_i p(\underline{n}) = \theta_i p(\underline{n} - \underline{1}_i) \qquad (3.167)$$

The formula equates the flow of probability flux out of state $\underline{n}$ from a departure from the ith queue to the flow into state $\underline{n}$ from state $\underline{n} - \underline{1}_i$ from an arrival to queue i.

This local balance equation can be written as

$$p(\underline{n}) = \frac{\theta_i}{\mu_i} p(\underline{n} - \underline{1}_i) \qquad (3.168)$$

or as

$$p(\underline{n}) = \frac{\theta_i}{\mu_i} p(n_1, n_2, \ldots, n_{i-1}, n_i - 1, n_{i+1}, \ldots, n_M) \qquad (3.169)$$

Utilizing the local balance equation repetitively, the ith term can be zeroed out

$$p(\underline{n}) = \left(\frac{\theta_i}{\mu_i}\right)^{n_i} p(n_1, n_2, \ldots, n_{i-1}, 0, n_{i+1}, \ldots, n_M) \qquad (3.170)$$

Then each term can be zeroed out

$$p(\underline{n}) = \left(\prod_{i=1}^{M} \left(\frac{\theta_i}{\mu_i}\right)^{n_i}\right) p(\underline{0}) \qquad (3.171)$$

The inverse of $p(\underline{0})$ is known as the normalization constant or $G(N)$. A physicist would call it a partition function. From the normalization equation

$$\sum_{\underline{n}} p(\underline{n}) = p(\underline{0}) G(N) = 1 \qquad (3.172)$$

$$G(N) = \sum_{\underline{n}} \left(\prod_{i=1}^{M} \left(\frac{\theta_i}{\mu_i} \right)^{n_i} \right) \tag{3.173}$$

The expression for $G(N)$ can be deduced from what we know of the form of the expression for $p(\underline{n})$ [equation (3.171)] and the normalization equation. The product form solution for closed Markovian queueing networks with random routing and ample buffers is then

$$p(\underline{n}) = \frac{1}{G(N)} \prod_{i=1}^{M} \left(\frac{\theta_i}{\mu_i} \right)^{n_i} \tag{3.174}$$

An open queueing network has an infinite number of states as it can have a potentially unlimited number of customers. A closed queueing network has a finite number of states as it has a finite number of customers. Again, the normalization constant can be thought of as a correction factor so that, if one sums over the finite (but possibly large) number of states of a closed Markovian queueing network, the sum of the state probabilities is one. Note $1/G(N)$ does not factor into $p_1(0)p_2(0)p_3(0)\ldots p_M(0)$ as it does for open networks, so closed networks are more difficult to solve. An algorithm for solving closed queueing networks is discussed in the next section.

A final thought that may occur to the reader is, if one has several incoming and outgoing transitions incident to a state, is there a simple way to know which transitions are paired for local balance? In fact the rule is that the flow of probability flux entering a state from an arrival to a specific queue balances with (equals) the flow of probability flux leaving the state from a departure from the same queue. This can be seen in equations (3.152) and (3.167).

3.6 Mean Value Analysis for Closed Networks

Even moderately sized queueing networks have a great many states. Consider a closed network of M queues and N customers. Finding the number of states is equivalent to finding the number of ways N identical customers can be in M queues. This number is:

$$\text{Number of States} = \binom{M + N - 1}{N} \tag{3.175}$$

As an example, suppose there are 8 queues and 60 customers. Then

$$\binom{67}{60} = 869,648,208 \text{ states} \tag{3.176}$$

Using the product form solution for closed networks, equation (3.174), would require the multiplication of nine constants to compute each of the state probabilities or a total of more than seven billion multiplications! For a larger closed networks or a network where there are customer classes the problem is even worse.

Is there a better way to compute closed Markovian (negative exponential service times) network performance measures? The answer is yes. The trick is not to calculate the state probabilities explicitly.

One approach, the convolution algorithm [see J. Buzen (Buzen 73) and M. Reiser and H. Kobayashi (Resier 73)], uses clever recursions to calculate the normalization constant G(N) [see equation (3.173)] and then performance measures (Bruell 80, Robertazzi 00). It is computationally efficient. Another computationally efficient approach is mean value analysis (MVA), from M. Reiser and S. Lavenberg (Reiser 80). The mean value analysis algorithm exactly computes each network queue's mean throughput, mean delay, and the mean number of customers, all through the use of some clever queueing principle-based recursions. Mean value analysis, unlike the convolution algorithm, does not compute normalization constants as part of its solution method. Two versions of the mean value algorithm are discussed below. Both involve state-independent servers. The first is for a cyclic network. In the second case the algorithm is generalized to a closed network with random routing.

3.6.1 MVA for Cyclic Networks

Again we have M Markovian queues and N customers. The network is cyclic. That is, the first queue's output is connected to the input of the second queue, the second queue's output is connected to the input of the third queue ... and the last queue's output is connected to the input of the first queue. The service rate of the ith queue is μ_i. The service time is negative exponentially distributed.

Now the average delay a customer undergoes at the ith queue $\overline{\tau_i}$ has two parts. One part is the delay at the server, and the second part is the sum of the service times of each customer in the queue ahead of the customer in question. Since the average service time of any customer is $1/\mu_i$ where μ_i is the service rate, one has

$$\overline{\tau_i} = \frac{1}{\mu_i} + \frac{1}{\mu_i} \times (\text{avg. number of customers in queue at arrival}) \quad (3.177)$$

The fact that this expression does not take into account the time that the customer in the server has been in service is from the memoryless nature of the negative exponential service time. That is, because the system is memoryless, the remaining time in service of the customer in the server of the ith queue at the time of a customer arrival always follows a negative exponential distribution with mean $1/\mu_i$.

Reiser and Lavenberg's insight was to realize (and prove) that in a closed Markovian network the number of customers in a queue a customer "sees" on its arrival to the queue has the same distribution as that for the network in equilibrium with one customer less. Thus

$$\overline{\tau_i}(N) = \frac{1}{\mu_i} + \frac{1}{\mu_i} \times \overline{n_i}(N-1) \tag{3.178}$$

In this equation $\overline{\tau_i}(N)$ is the average delay for the ith queue when the network has N customers. Also, $\overline{n_i}(N)$ is the average number of customers in the ith queue when there are N customers in the network.

For the rest of the mean value analysis algorithm, one can use Little's Law [see equation (3.32)]. First, applying Little's Law to the entire cyclic network,

$$\overline{T}(N) \sum_{i=1}^{M} \overline{\tau_i}(N) = N \tag{3.179}$$

or

$$\overline{T}(N) = \frac{N}{\sum_{i=1}^{M} \overline{\tau_i}(N)} \tag{3.180}$$

Here $\overline{T}(N)$ is the average throughput in the network. Since the queues are arranged in cyclic fashion, $\overline{T}(N)$ is the same for every queue and is equal to each queue's arrival rate.

Second, for the ith queue

$$\overline{n_i}(N) = \overline{T}(N)\overline{\tau_i}(N) \tag{3.181}$$

Starting with $\overline{n_i}(0) = 0$ for $i = 1, 2 \ldots M$ one can use equations (3.178), (3.180), and (3.181) to recursively compute the mean delay, the mean throughput, and the mean number of customers in each queue. One has for the mean value analysis algorithm for cyclic networks

MVA Algorithm for Cyclic Networks

For $i = 1, 2, 3 \ldots M$

$$\overline{n_i}(0) = 0 \tag{3.182}$$

For $N = 1, 2, 3$

   For $i = 1, 2, 3 \ldots M$

$$\overline{\tau_i}(N) = \frac{1}{\mu_i} + \frac{1}{\mu_i} \times \overline{n_i}(N-1) \tag{3.183}$$

$$\overline{T}(N) = \frac{N}{\sum_{j=1}^{M} \overline{\tau_j}(N)} \tag{3.184}$$

$$\overline{n_i}(N) = \overline{T}(N)\overline{\tau_i}(N) \tag{3.185}$$

Example: M Identical Cyclic Queues

This canonical example appears in Schwartz 87. There are M cyclic queues all with service rate μ. Since all of the queues have the same service rate, the mean number of customers, the mean delay, and the mean throughput for each queue are the same.

Now with one customer and M queues

$$\overline{n_i}(0) = 0 \tag{3.186}$$

$$\overline{\tau_i}(1) = \frac{1}{\mu} \tag{3.187}$$

$$\overline{T}(1) = \frac{1}{M\frac{1}{\mu}} = \frac{\mu}{M} \tag{3.188}$$

$$\overline{n_i}(1) = \frac{\mu}{M} \times \frac{1}{\mu} = \frac{1}{M} \tag{3.189}$$

With two customers and M queues

$$\overline{\tau_i}(2) = \frac{1}{\mu} + \left(\frac{1}{\mu} \times \frac{1}{M} \right) = \frac{1}{\mu} \left(\frac{M+1}{M} \right) \tag{3.190}$$

$$\overline{T}(2) = \frac{2}{M \times \frac{1}{\mu} \times \left(\frac{M+1}{M} \right)} = \frac{2\mu}{M+1} \tag{3.191}$$

$$\overline{n_i}(2) = \frac{2\mu}{M+1} \times \frac{1}{\mu} \times \left(\frac{M+1}{M} \right) = \frac{2}{M} \tag{3.192}$$

With three customers and M queues

$$\overline{\tau_i}(3) = \frac{1}{\mu} + \left(\frac{1}{\mu} \times \frac{2}{M} \right) = \frac{1}{\mu} \left(\frac{M+2}{M} \right) \tag{3.193}$$

$$\overline{T}(3) = \frac{3}{M \times \frac{1}{\mu} \times \left(\frac{M+2}{M} \right)} = \frac{3\mu}{M+2} \tag{3.194}$$

$$\overline{n_i}(3) = \frac{3\mu}{M+2} \times \frac{1}{\mu} \times \left(\frac{M+2}{M} \right) = \frac{3}{M} \tag{3.195}$$

A pattern can be observed in these results. The general solution for N customers over M queues is

$$\overline{\tau_i}(N) = \frac{1}{\mu} \left(\frac{M + N - 1}{M} \right) \tag{3.196}$$

$$\overline{T}(N) = \frac{N\mu}{M + N - 1} \tag{3.197}$$

$$\overline{n_i}(N) = \frac{N}{M} \tag{3.198}$$

3.6.2 MVA for Random Routing Networks

In this more general case, the network allows random routing between the queues. Again, the independent probability that a customer leaving the ith queue enters the jth queue is r_{ij}.

Now the θ_i's, the solutions of the traffic equations (3.166), have to be taken into account. We choose (the nonunique) θ_i's so that one queue's average throughput is 1.0. This is the reference queue. Then:

MVA Algorithm for Random Routing Networks
For $i = 1, 2, 3 \dots M$

$$\overline{n_i}(0) = 0 \tag{3.199}$$

For $N = 1, 2, 3$
*   For $i = 1, 2, 3 \dots M$*

$$\overline{\tau_i}(N) = \frac{1}{\mu_i} + \frac{1}{\mu_i} \times \overline{n_i}(N - 1) \tag{3.200}$$

$$\overline{T}(N) = \frac{N}{\sum_{j=1}^{M} \theta_j \overline{\tau_j}(N)} \tag{3.201}$$

$$\overline{n_i}(N) = \overline{T}(N)\theta_i \overline{\tau_i}(N) \tag{3.202}$$

Here $\overline{T}(N)$ is the actual unique throughput of the reference queue when there are N customers. The mean throughput of the ith, nonreference queue is $\theta_i \overline{T}(N)$. For a cyclic network, $\theta_i = 1$ for all i and the above equations simplifies to the earlier equations.

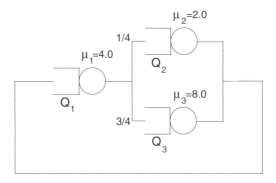

Fig. 3.15. A random routing network. Here $\mu_1 = 4.0$, $\mu_2 = 2.0$, $\mu_3 = 8.0$, $r_{12} = 0.25$, and $r_{13} = 0.75$

Example: Three Queues with Random Routing

Let us run the mean value analysis algorithm for a closed Markovian queueing network with random routing for the three-queue network illustrated in Figure 3.15. The service rates of Q1, Q2, and Q3 are 4.0, 2.0, and 8.0, respectively. The reference queue is queue 1. The routing probabilities are $r_{12} = 0.25$ and $r_{13} = 0.75$. Therefore the relative throughputs are $\theta_1 = 1.0$, $\theta_2 = 0.25$, and $\theta_3 = 0.75$.

With one customer for this queueing network

$$\overline{n_i}(0) = 0 \qquad\qquad i = 1, 2, 3 \qquad\qquad (3.203)$$

$$\overline{\tau_1}(1) = 0.25 \qquad\qquad (3.204)$$

$$\overline{\tau_2}(1) = 0.50 \qquad\qquad (3.205)$$

$$\overline{\tau_3}(1) = 0.125 \qquad\qquad (3.206)$$

$$\overline{T}(1) = \frac{1}{1.0 \times 0.25 + 0.25 \times 0.50 + 0.75 \times 0.125} = 2.1333 \qquad (3.207)$$

$$\overline{n_1}(1) = 2.1333 \times 1.0 \times 0.25 = 0.53333 \qquad\qquad (3.208)$$

$$\overline{n_2}(1) = 2.1333 \times 0.25 \times 0.50 = 0.26666 \qquad\qquad (3.209)$$

$$\overline{n_3}(1) = 2.1333 \times 0.75 \times 0.125 = 0.20000 \qquad\qquad (3.210)$$

With two customers

$$\overline{\tau_1}(2) = 0.25 + 0.25 \times 0.5333 = 0.38333 \tag{3.211}$$

$$\overline{\tau_2}(2) = 0.50 + 0.50 \times 0.2666 = 0.63333 \tag{3.212}$$

$$\overline{\tau_3}(2) = 0.125 + 0.125 \times 0.2 = 0.15 \tag{3.213}$$

$$\overline{T}(2) = \frac{2}{1.0 \times 0.38333 + 0.25 \times 0.63333 + 0.75 \times 0.15} = 3.05732 \tag{3.214}$$

$$\overline{n_1}(2) = 3.05732 \times 1.0 \times 0.38333 = 1.17197 \tag{3.215}$$

$$\overline{n_2}(2) = 3.05732 \times 0.25 \times 0.63333 = 0.484076 \tag{3.216}$$

$$\overline{n_3}(2) = 3.05732 \times 0.75 \times 0.15 = 0.343949 \tag{3.217}$$

With three customers

$$\overline{\tau_1}(3) = 0.25 + 0.25 \times 1.17197 = 0.54299 \tag{3.218}$$

$$\overline{\tau_2}(3) = 0.50 + 0.50 \times 0.484076 = 0.74204 \tag{3.219}$$

$$\overline{\tau_3}(3) = 0.125 + 0.125 \times 0.343949 = 0.16799 \tag{3.220}$$

$$\overline{T}(3) = \frac{3}{1.0 \times 0.54299 + 0.25 \times 0.74204 + 0.75 \times 0.16799} = 3.51085 \tag{3.221}$$

$$\overline{n_1}(3) = 3.51085 \times 1.0 \times 0.54299 = 1.9064 \tag{3.222}$$

$$\overline{n_2}(3) = 3.51085 \times 0.25 \times 0.74204 = 0.65130 \tag{3.223}$$

$$\overline{n_3}(3) = 3.51085 \times 0.75 \times 0.16799 = 0.44234 \tag{3.224}$$

3.7 Negative Customer Queueing Networks

There are applications where it would be useful to have a queueing model where customers can "disappear" from a network. It may be desired for a customer to disappear if older messages are canceled by newer ones or if one is modeling a real-time system.

Models with "negative customers" do model this type of activity. In such models, there are normal, "positive" customers and negative customers. A negative customer arriving to a queue will cancel a positive customer. That is, both will instantly disappear from the system. Much in the spirit of matter and anti-matter, positive and negative customers annihilate each other.

It should be noted that the application that E.Gelenbe, the original creator of negative customer models (Gelenbe 91a, 91b) had in mind was neural network modeling. With a neuron modeled by a queue, positive customers represent excitation signals and negative customers represent inhibition signals.

How is the generation of negative customers modeled? Positive customers arrive to the ith queue according to a Poisson process with mean arrival rate Λ_i. Negative customers arrive to the ith queue according to a Poisson process with mean λ_i.

A second way that a negative customer may be generated is as a queue departure. That is, a positive customer leaving the ith queue enters the jth queue as a negative customer with independent probability r_{ij}^- and enters the jth queue as a positive customer with independent probability r_{ij}^+. Also, a positive customer leaving the ith queue departs from the network with probability d_i. Finally, the service rate of customers in the ith queue is independent and negative exponentially distributed with mean service rate μ_i. Note that a negative customer arriving to an empty queue instantly disappears from the network.

Equilibrium results for negative customer networks only make sense in the context of open networks, as for closed networks, the network would be empty in a finite amount of time as all positive customers would eventually be destroyed. It should also be realized that the model described above is Markovian (memoryless).

It turns out that Gelenbe and co-authors developed a negative customer network model that has a product form solution. This model will now be presented. An alternative model from Chao and Pinedo appears in Chao 93. (see also Robertazzi 00). A survey of work on negative customer models is Artalejo 00.

3.7.1 Negative Customer Product Form Solution

Let's define the effective utilization of the ith queue as

$$q_i = \frac{\lambda_i^+}{\mu_i + \lambda_i^-} \tag{3.225}$$

The λ_i^+ and λ_i^- here are the solutions of the following traffic equations

$$\lambda_i^+ = \sum_j q_j \mu_j r_{ji}^+ + \Lambda_i \tag{3.226}$$

$$\lambda_i^- = \sum_j q_j \mu_j r_{ji}^- + \lambda_i \qquad (3.227)$$

Note that, if the definition of the q_i is substituted into the traffic equations, it can be seen that they are nonlinear, unlike the linear traffic equations of networks with only positive customers (see section 3.5). Here λ_i^+ is the mean arrival rate of positive customers into the ith queue from both queue transfers (first term) and external arrivals (second term). Similarly λ_i^- is the mean arrival rate of negative customers to the ith queue from queue transfers and external arrivals. From the definition of q_i, it can be seen that the flow of arriving negative customers to a queue increases its effective service rate ($\mu_i + \lambda_i^-$).

Naturally from above

$$\sum_j r_{ij}^+ + \sum_j r_{ij}^- + d_i = 1 \qquad 1 \le i \le M \qquad (3.228)$$

That is, a customer departing the ith queue either enters another jth queue as a positive or a negative customer or departs from the system. There are M queues in the network.

To solve for an expression for the equilibrium state probabilities, one needs to first set up a global balance equation for a state. Some notation in terms of the number of positive customers needed in each queue is

$$\underline{n} = (n_1, n_2, \ldots n_i \ldots \ldots n_M) \qquad (3.229)$$

$$\underline{n}_i^+ = (n_1, n_2, \ldots n_{i-1}, n_i + 1, n_{i+1}, \ldots n_M) \qquad (3.230)$$

$$\underline{n}_i^- = (n_1, n_2, \ldots n_{i-1}, n_i - 1, n_{i+1}, \ldots n_M) \qquad (3.231)$$

$$\underline{n}_{ij}^{+-} = (n_1, n_2, \ldots n_i + 1, \ldots n_j - 1, \ldots n_M) \qquad (3.232)$$

$$\underline{n}_{ij}^{++} = (n_1, n_2, \ldots n_i + 1, \ldots n_j + 1, \ldots n_M) \qquad (3.233)$$

The first vector simply represents the state $\underline{n}$. The second vector indicates the network population after an external positive customer arrival to the ith queue while the network is in state $\underline{n}$. The third vector indicates the network population after a departure from queue i while the network is in state $\underline{n}$. The fourth vector corresponds to the state prior to a positive customer departure from queue i to queue j that brings the state to state $\underline{n}$. Finally, the last vector for $\underline{n}_{ij}^{++}$ corresponds to the state prior to a negative customer leaving the ith queue for the jth queue and bringing the network to state $\underline{n}$. Note that in all of this negative customers have lifetimes of 0 seconds as once created they instantly cancel a positive customer and/or disappear from the network.

Also let $1[y]$ be an indicator function that has a value of 1.0 if y is greater than zero and has a value of zero otherwise.

Then the global balance equation for an interior state is

$$p(\underline{n}) \sum_i [\Lambda_i + (\lambda_i + \mu_i)1[n_i > 0]] \tag{3.234}$$

$$= \sum_i [p(\underline{n}_i^+)\mu_i d_i + p(\underline{n}_i^-)\Lambda_i 1[n_i > 0] + p(\underline{n}_i^+)\lambda_i$$

$$+ \sum_j (p(\underline{n}_{ij}^{+-})\mu_i r_{ij}^+ 1[n_j > 0] + p(\underline{n}_{ij}^{++})\mu_i r_{ij}^- + p(\underline{n}_i^+)\mu_i r_{ij}^- 1[n_j = 0])]$$

Gelenbe found that the expression for the equilibrium state probability $p(\underline{n})$ that satisfies this global balance equation is

$$p(\underline{n}) = \prod_{i=1}^M (1 - q_i){q_i}^{n_i} \tag{3.235}$$

It can be seen that this expression is similar in form to the earlier one of section 3.5.1 for open Markovian networks with only positive customers. That the equilibrium state probabilities has such a product form is not at all an obvious result.

Example: Tandem Network

Consider a tandem (series) of two queues, queue 1 followed by queue 2 (Figure 3.16). Positive customers arrive to queue 1 with arrival rate Λ_1. No negative customers arrive to queue 1.

There is a Poisson stream of external negative customer arrivals to queue 2 with mean rate λ_2. Also, a customer departing queue 1 enters queue 2 as a positive customer with probability r_{12}^+ and as a negative customer with probability r_{12}^-. The service rates of queues 1 and 2 are μ_1 and μ_2, respectively.

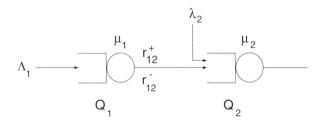

Fig. 3.16. A negative queueing network example

The traffic equations are

$$\lambda_1^+ = \Lambda_1 \tag{3.236}$$

$$\lambda_1^- = 0 \tag{3.237}$$

$$\lambda_2^+ = q_1 \mu_1 r_{12}^+ \tag{3.238}$$

$$\lambda_2^- = q_1 \mu_1 r_{12}^- + \lambda_2 \tag{3.239}$$

Then from equation (3.225)

$$q_1 = \frac{\Lambda_1}{\mu_1} \tag{3.240}$$

$$q_2 = \frac{q_1 \mu_1 r_{12}^+}{\mu_2 + q_1 \mu_1 r_{12}^- + \lambda_2} \tag{3.241}$$

or

$$q_2 = \frac{\Lambda_1 r_{12}^+}{\mu_2 + \Lambda_1 r_{12}^- + \lambda_2} \tag{3.242}$$

where

$$p(\underline{n}) = \prod_{i=1}^{2} (1 - q_i) q_i^{n_i} \tag{3.243}$$

It can be observed that queue 1 is solely a positive customer queue. Thus queue 1 has the standard positive customer queue utilization.

Networks of queues with negative customers have been generalized over the years to such instances as networks where the customers may leave in batches, to networks with disasters (a single arriving negative customer causes all positive customers in a queue to be removed from the network), and to multiple classes of positive and negative customers. See Artalejo 00 for a review.

3.8 Recursive Solutions for State Probabilities

Not every queueing model of interest is a product form network. Non-product form models are defined by exclusion: They are models that do not have a product form solution. Such realistic queueing features, as blocking, priority classes, and finite buffers gives rise to non-product form models. Any non-product form model may be solved, if it is not too large, by solving the model's

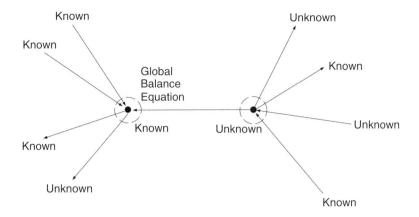

Fig. 3.17. Type A state transition diagram structure

global balance equations. However some non-product form queueing models can be efficiently solved recursively for the equilibrium state probabilities. This was first discussed by Herzog, Woo, and Chandy in 1975.

There are three ways that one can create recursions to generalize non-product form model equilibrium state probabilities. One is to draw boundaries that segment the state transition diagram into two parts. One can then write equations that balance the flow of probability flux moving across the boundary in both directions.

The other two methods for generating recursions involve two ways of writing global balance equations (Wang 90). Figure 3.17 illustrates a "type A" structure in the state transition diagram. Here one can write a global balance equation for a state with (previously found) known state probability where there is only one incident transition to the state from a state with unknown state probability. In solving the global balance equation, the unknown state probability is found. One may then continue by solving the global balance equation for that state with (now) known state probability.

For "type B" structure (Figure 3.18) one writes a global balance equation for a state with unknown state probability, with incident transitions from states with known probabilities and with departing transitions to states with known and/or unknown state probabilities. Once one solves the global balance equation for its single unknown state probability, one can move on to another state with unknown probability connected via an outgoing transition and continue the process. Note that states with unknown probability reached by departing transitions do not have their state probabilities enter into the global balance equation calculation.

It is possible to generate these type A and B recursions from subsets of states at a time, rather than from single states. A more detailed discussion of such models appears in Robertazzi 00. An example is now presented.

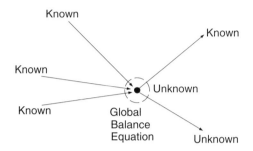

Fig. 3.18. Type B state transition diagram structure

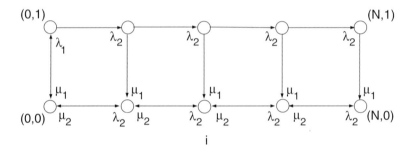

Fig. 3.19. Integrated voice/data protocol recursive solution state transition diagram

Example: Voice/Data Integrated Protocol

Let's consider a link that can either carry a single voice call or packet transmissions, but not both at the same time. The continuous time state transition diagram appears in Figure 3.19. The horizontal axis indicates the number of packets in the transmission buffer. The buffer size is N. The vertical axis indicates either no voice call present (0) or a single voice call present (1).

A voice call arrives with Poisson rate λ_1 and completes service in a negative exponentially distributed amount of time with service rate μ_1. Packets arrive with Poisson rate λ_2, and each completes service in a negative exponentially distributed amount of time with service rate μ_2.

In this protocol one of several packets at a time is only transmitted if there is no voice call (Schwartz 87). Otherwise the packets are buffered until the voice call finishes. A voice call is only accepted if the packet buffer is empty.

Let $p(0,0) = 1.0$. The state probabilities can be normalized when the recursions are finished. A global balance equation at state $(0,1)$ is

$$(\lambda_2 + \mu_1)p(0,1) = \lambda_1 p(0,0) \tag{3.244}$$

So

$$p(0,1) = \frac{\lambda_1}{\lambda_2 + \mu_1}p(0,0) \tag{3.245}$$

Moving from left to right through the state transition diagram, bottom row equilibrium state probabilities can be calculated by drawing vertical boundaries through the state transition diagram and equating the flow of probability flux from left to right to that from right to left.

$$\mu_2 p(i,0) = \lambda_2[p(i-1,0) + p(i-1,1)] \tag{3.246}$$

or

$$p(i,0) = \frac{\lambda_2}{\mu_2}[p(i-1,0) + p(i-1,1)] \tag{3.247}$$

Top row states follow a type B structure. Their equilibrium state probabilities can be solved from left to right by writing the global balance equation

$$(\lambda_2 + \mu_1)p(i,1) = \lambda_2 p(i-1,1) \tag{3.248}$$

or

$$p(i,1) = \frac{\lambda_2}{\lambda_2 + \mu_1}p(i-1,1) \tag{3.249}$$

After computing p(0,0) and p(0,1), one can cycle through equations (3.247) and (3.249), where $i = 1,2,3\ldots N-1$.

Finally, at the right boundary using global balance equations for states $(N,0)$ and $(N,1)$ and some simple algebra results in

$$p(N,0) = \frac{\lambda_2}{\mu_2}[p(N-1,0) + p[(N-1,1)] \tag{3.250}$$

$$p(N,1) = \frac{\lambda_2}{\mu_1}p(N-1,1) \tag{3.251}$$

Thus the above can be used to solve the state probabilities in terms of reference probability $p(0,0)$. The probabilities can then be normalized by dividing each by the sum of the unnormalized probability values.

3.9 Stochastic Petri Nets

Petri networks are a graphical means of representing serialization, concurrency, resource sharing, and synchronization. Stochastic Petri networks use stochastic timing for events. Markovian Petri networks use memoryless distributions for timing. Just as Markovian queueing network schematic diagrams give rise to Markov chains, Markovian Petri network schematic diagrams give rise to Markov chains.

Petri networks first appeared in the doctoral dissertation of C. A Petri in Germany in 1962.

3.9.1 Petri Net Schematics

Stochastic Petri network schematics consist of a six tuple.

$$\underline{P} = (P, T, I, O, M, Q) \tag{3.252}$$

As an example suppose we have two processors (P1 and P2) connected through a computer bus to a common memory (CM). Only one processor may use the bus to access the common memory at a time. A Petri network schematic for common memory access appears in Figure 3.20. More details on such multiprocessors appear in Marsan 83,86.

Referring to the diagram, P is a set of "places" that are drawn as circles. We use T to represent a set of "transitions" that are drawn as horizontal bars. The input function I maps each transition to one or more places. The output function O maps each place to one or more transitions. The input and output functions are represented by directed arcs. A "marking" M assigns zero or more tokens (illustrated by dots) to each place. A specific marking is a state of the Petri net. Finally, Q is the set of transition rates associated with the transitions.

How does the Petri net schematic operate? When there is at least one token in each place incident to a transition, the transition is "enabled." The transition can then "fire" after some period of time. Firing involves removing one token from each place that is incident to the transition of interest and adding a token to each place that outgoing arcs lead to from the transition. Naturally the firing of a transition, by changing the marking, leads to a new network state.

In a Markovian Petri net (often called a "stochastic" Petri net), the time between when a transition is enabled for firing and when it actually fires is an independent negative exponentially distributed random variable. Such transitions are drawn as unfilled rectangles as in the figure. Immediate transitions fire in zero time once enabled and are drawn as filled rectangles. Generalized stochastic Petri networks contain both immediate and negative exponential timed transitions. All of the transitions in Figure 3.20 have Markovian timing.

In Figure 3.20 the marking indicates that P1 and P2 are idle (not trying to connect to the common memory) and the bus is free. The sequence of places

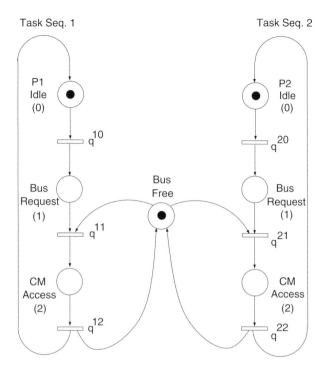

Fig. 3.20. A stochastic Petri net for common memory access from two processors over a bus

Pi idle, Bus request, and CM access comprise a linear task sequence. Either or both of the linear task sequences can proceed through the firing of the q^{10} and q^{20} transitions to requesting the bus. Since a processor accessing the bus removes the bus free place token while it accesses the memory, only one processor can access the common memory (bus) at a time. Once a common memory access is finished the accessing processor becomes idle and the bus becomes free again.

This Petri net can be seen to model concurrency in having two parallel task sequences, and to model serializability in the serial nature of each task sequence. Resource sharing is modeled through the bus free place. Finally, by requiring that transitions q^{11} and q^{21} fire only when there is a bus request and the bus is free, synchronization is modeled.

3.9.2 Petri Net Markov Chains

The regular structure of the Petri net schematic of Figure 3.20 lends itself to creating a corresponding Markov chain that is Cartesian coordinate based. The Markov chain is drawn in Figure 3.21. The horizontal axis corresponds

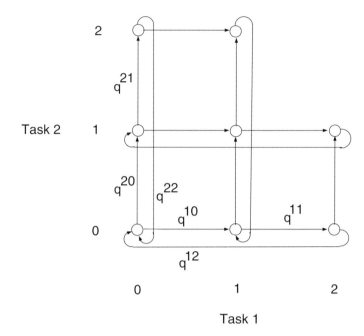

Fig. 3.21. Markov chain of the Petri net example

to task sequence 1, and the vertical axis corresponds to task sequence 2. The coordinates 0, 1, and 2 represent a token being in the P_i idle, Bus request, and CM access places, respectively. Note that the "wraparound" character of the Markov chain is naturally embedded on the surface of a torus (Robertazzi 00).

Stochastic Petri nets can be solved through simulation or by Markov chain solution. The Markov chain in the figure is a non-product form Markov chain. One can solve the set of global balance equations for the equilibrium state probabilities.

However, if transitions [(2,0) to (2,1)] and [(0,2) to (1,2)] are removed from the chain, a different protocol results, which has a product form solution. The corresponding Petri net is shown in Figure 3.22. The inhibitor arcs that have been added implement a complementary dependency. That is, the condition for the transition that the inhibitor arcs are attached to, to fire, includes there being no token in the place from which the inhibitor arc originates. In the figure, the inhibitor arc in task sequence 1 is connected to the CM access place in task sequence 2 ("T.S. 2") and vice versa. Thus one can only move from a processor being idle to a bus request if there is a nonzero probability that one can cycle completely around the processors' task sequence without the need for a state (marking) change in the other task sequence. This precludes a form of blocking (a bus request not being satisfied because the other processor is

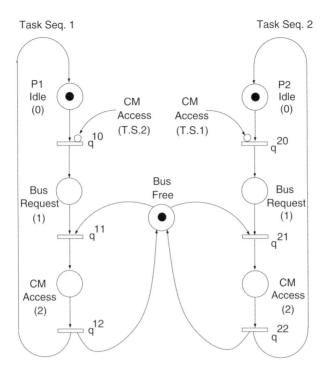

Fig. 3.22. Modified Petri net with product form solution

currently accessing the common memory) that does not allow a product form solution.

The global balance equation for the modified Markov chain is

$$(q^{1,i} + q^{2,j})p(i,j) = q^{1,i-1}p(i-1,j) + q^{2,j-1}p(i,j-1) \qquad (3.253)$$

Here $q^{k,l}$ is the transition rate of the lth Petri schematic transition in the kth task sequence. Also, $p(i,j)$ is the equilibrium state probability of the (i,j)th state in the state transition diagram.

For the local balance equations, one has

$$q^{1,i-1}p(i-1,j) = q^{1,i}p(i,j) \qquad (3.254)$$

$$q^{2,j-1}p(i,j-1) = q^{2,j}p(i,j) \qquad (3.255)$$

The product form solution for the Markov chain without the two deleted transitions is

$$p(i,j) = \frac{q^{1,0}}{q^{1,i}} \frac{q^{2,0}}{q^{2,j}} p(0,0) \qquad (3.256)$$

This product form solution satisfies both the global and the local balance equations.

Petri nets are a flexible tool for system modeling. More on stochastic Petri network modeling appears in Robertazzi 00.

3.10 Solution Techniques

A variety of solution techniques are available for solving queueing and stochastic Petri network models of networks (and computer systems). Each technique has advantages and disadvantages in terms of its modeling ability, ease of implementation, and ease of use.

3.10.1 Analytical Solutions

Analytical solutions involve developing a mathematical model that can be solved to produce a closed form formula yielding the desired result. In fact the formula may be solved on a computer (as in the Erlang B or C formula), but usually the amount of computation required is trivial.

Because of the simplicity and the intuitive insight they offer, analytical closed form solutions are the most desirable of solutions. Unfortunately analytical solutions tend to be available only for simpler models. Although it is sometimes possible to derive analytical solutions for more complex models, often the skills of a highly educated Ph.D. are required to find the solution.

3.10.2 Numerical Computation

Beyond implementing a simple formula on a computer (or even calculator), some approaches involve the numerical solution of mathematical equations modeling a system. Three examples are as follows:

- *Linear Equation Solution:* In theory any Markovian system can be exactly solved for its equilibrium state probabilities by solving its linear global balance equations. As mentioned, N states give rise to N equations with one of them replaced by the normalization equation to obtain a set of N equations with a unique and correct solution. The difficulty, of course, is that even moderate size systems have so many states (and thus equations) as to make this not a computationally feasible approach. Moreover the fact that the computational complexity of general linear equation solution is proportional to the cube of the number of equations compounds this problem.

- *Transient Models:* Transient models involve the operation of a system over a limited time span (say from 0 to 2 seconds). Thus this is not a model in equilibrium. In theory, standard positive customer memoryless queueing networks can be solved in continuous time by coupled linear differential equations and in discrete time by coupled linear difference equations. Software packages for such systems of equations are available, although their use is only practical for smaller systems.

• *Generating Functions:* As has been stated, moment generating functions provide a frequency domain-like representation of equilibrium state probabilities. The moment generating function of a linear set of equilibrium state probabilities is

$$P(z) = \sum_{n=0}^{\infty} p_n z^n \qquad (3.257)$$

This expression can be seen to be similar, although not identical, to the z transform of digital signal processing. Producing a closed form moment generating function expression is an exercise in analysis. However, such expressions, which are functions of z, can be numerically inverted on a computer to produce the equilibrium state probabilities.

Today it is of more interest than ever to produce complete probability distributions, rather than simply low-order moments, because of the interest in low probability events such as the overflow of buffers (Michiel 97).

3.10.3 Simulation

Simulation is a middle approach between mathematical models and experimentation to determine a system's performance. In a discrete event simulation, a program mimics the actual system (e.g., calls are initiated and terminated, packets transmitted, and buffers overflow) along with the timing of events. Part of the program collects statistics on the operation of the modeled system, which are synthesized into performance results.

Simulation is cost effective in capturing realistic modeling features (e.g., blocking in networks, priority classes, and non-Markovian statistics) that preclude analytical or even numerical solutions. Generally one does not have to be as sophisticated mathematically to produce a simulation as to produce analytical solutions. Both steady state and transient system operation can be simulated.

The size of a system to be simulated is limited by the available computer power. Very large systems such as the Internet cannot be simulated in extreme detail. Moreover, since simulations can produce voluminous performance results, it is often easier to discern systematic trends and trade-offs with analytic solutions.

Naturally a simulation of a queueing or stochastic Petri network involves many random quantities. These are generated by a "pseudo-random" number generator. These software programs generate random-like sequences of numbers that follow the same sequence for a given "seed" number. Moreover, after a very large number of pseudo-random numbers, the sequence repeats.

Although random-like, the sequence is really deterministic. This aids in reproducing results. That is, for the same seed and no parameter changes, a simulation will produce exactly the same results every time it is run. Changing the seed produces statistically similar, although not identical, results.

Pseudo-random numbers are usually uniformly distributed between 0 and 1. If one needs the probability of a packet arrival in a slot to be 0.2, there is an arrival if the pseudo-random number is between 0.0 and 0.2 and there is no arrival if the pseudo-random number is between 0.2 and 1.0.

How does one generate random numbers following non-uniform distributions? Say $f(x)$ is the continuous probability distribution we want. Then $F(x)$ is the cumulative distribution function (Goodman 04).

$$F(x) = \int_{-\infty}^{x} f(z)dz \qquad (3.258)$$

Here z is a dummy integration variable. To generate a random variable with distribution $f(x)$, one takes a pseudo-random uniformly distributed variable y and lets

$$x = F^{-1}(y) \qquad (3.259)$$

The function $F^{-1}(y)$ is the functional inverse of the cumulative distribution function. For each uniformly distributed y, an x is generated according to the above formula that follows the distribution $f(x)$. An analogous procedure can create discrete random variables of any distribution.

Often a simulation is run many times, each time with a different seed, and the results are averaged. Confidence intervals provide a means of expressing the amount of variability in such results. A 98% confidence interval consists of an interval with two end points such that 98% of the time the quantity of interest is within the interval.

In considering confidence intervals (e.g., 90%, 95%, 98%, and 99%), one should realize that the smaller the percentage, the tighter the upper and lower limits are and the smaller the confidence interval is. Confidence intervals are often plotted as vertical lines (bars) superimposed on performance curves where the length of the line expresses the size of the confidence interval at each data point. The smaller the bars, the smaller is the variability in the curve.

Finally a sensitivity analysis determines the degree to which performance changes if a parameter value changes slightly.

3.11 Conclusion

Queueing theory and stochastic Petri network theory cover intriguing problems that have captured the imagination of many researchers and developers.

The ubiquity of calls and packets waiting in buffers and of concurrency, serialization, synchronization, and resource sharing means that such problems will be of interest for some time to come.

3.12 Problems

1. What are the statistical assumptions behind an M/M/1 queue? A Geom/Geom/1 queue?
2. What does it mean to say that a M/M/1 queue is memoryless?
3. Compare Markov chains with electric circuits.
4. Give an example of the use of Little's Law.
5. Describe the difference between local and global balance.
6. What is the computational problem with solving global balance equations for large Markov chains?
7. If a M/M/1 queue input is Poisson, what can one say about the output process?
8. Can the arrival rate be greater than the service rate for a M/M/1 finite buffer queue? Explain.
9. Explain in which situation the use of the Erlang B formula is appropriate. Do the same for the Erlang C formula.
10. Give and explain an application for the queueing-based memory model of section 3.3.4.
11. What does it mean to use a Markov chain "embedded" at departure instants for the M/G/1 queue analysis?
12. What are the statistical assumptions behind the Markov queueing results of section 3.5?
13. What do the traffic equations of section 3.5 model?
14. What queueing relationship is the mean value analysis algorithm based on? Does the MVA algorithm compute normalization constants?
15. What happens when a negative customer enters a queue with at least one positive customer? What happens when a negative customer enters an empty queue?
16. Can non-product form queueing networks be simply solved? Always or sometimes?
17. What types of actions do Petri nets model?
18. How is a Petri net marking related to its state?
19. What is the advantage of analytical studies of queueing and stochastic Petri nets?
20. Will simulation results always match experimental results?
21. Name three types of models discussed in this chapter having product form solutions.
22. Consider the Markov chain of Figure 3.23 with three classes of packets and a buffer that holds one packet at one time.

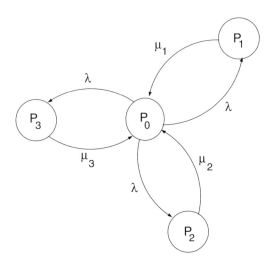

Fig. 3.23. A specific Markov chain

(a) Using boundaries, find p_1, p_2, and p_3 as functions of p_0, λ, and μ. Also find an expression for p_0.

(b) Find the blocking probability (i.e., the probability an arriving customer is turned away).

23. Consider a finite buffer Geom/Geom/1/N discrete time queue. Assume that a customer arriving to an empty queue must wait at least one slot for service. Assume also (particularly if the buffer is full) that departures occur before arrivals in a slot.

(a) Draw and label the state transition diagram.

(b) Solve for P_n in terms of p, s, and P_0 and solve for P_0.

(c) Solve for P_0 if an infinite size buffer is used.

24. Consider a discrete time Geom/Geom/2/4 queue (two servers holding at most four customers). A packet can enter an empty queue and depart during the same slot. Departures occur before arrivals in a slot. Draw and carefully label the state transition diagram.

25. Consider a single server finite buffer queue with the following state probabilities:

n	$p(n)$
0	0.15
1	0.20
2	0.35
3	0.30

Find the mean number of customers, the mean throughput, and the mean delay through the system.

26. Consider a finite buffer M/M/1/N queue where the number of customers N is 3. Let $\lambda = 3.0$ and $\mu = 4.0$.

(a) Draw and label the state transition diagram and calculate, numerically, p_0, p_1, p_2, and p_3.

(b) Calculate the average (mean) delay through the queue (for customers that are not blocked, of course).

(c) What is the average time that a customer waits in the waiting line before entering the server?

27. Consider an M/M/1 queue with $\lambda = 2.0$ and $\mu = 7.0$.

(a) If the buffer is infinite in size, find the numerical value of utilization.

(b) If now one has a finite buffer queue with the same parameters and a maximum capacity of $N = 4$, find the numerical utilization.

(c) One of the values of the previous two parts is greater. Intuitively, why is this so?

28. Consider an M/M/2/4 queueing system with two servers and a maximum capacity of four customers. Find the blocking probability formula for arriving customers. It is a function of λ and μ. Note: The answer is different from the Erlang B formula although the technique to find it is similar.

29. Consider a Markovian queueing system of three parallel servers without a queue. The arrival rate to the system of queues is $\lambda = 10.0$, and the service rate μ of each server is 6.0.

(a) Write an expression for the average number of empty servers as a function of the state probabilities.

(b) Calculate numerical values for the equilibrium state probabilities. Show all steps in doing this. Substitute these probabilities into the answer of (a) to find a numerical value for the average number of empty servers.

30. A small company has three outside telephone lines. An average call lasts 12 minutes, and 9 calls per hour are generated. Calls that do not immediately get a telephone line are queued (the network rings the caller when a line is available under a FIFO discipline).

(a) What queuing model is described above?

(b) What formula can be used to find the probability of queueing?

31. Consider a D/D/1 queue where one customer arrives per second and the server can process two customers a second. The arrival process and service process is deterministic (the time between events is a constant). Assuming that the queue is empty at $t = 0$ seconds, sketch the number of customers in the queue versus time for 0 to 3 seconds (label the graph accurately). From the graph and intuition, what is the average number of customers in the queue over an extended period of time?

32. For an M/M/∞ queueing system (where every arriving customer gets its own server of rate μ), what is the throughput? Use intuition, rather than calculation, to answer this question.

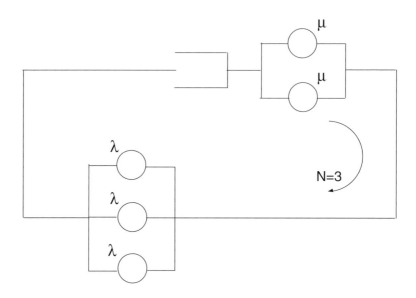

Fig. 3.24. A cyclic queueing system

33. In a queue with discouraged arrivals as the waiting line gets longer, fewer new customers enter the queue. Assume that for a Markovian queue the service rate is μ and the state dependent arrival rate is

$$\lambda(n) = \frac{\lambda}{n+1} \qquad n = 0, 1, 2, 3 \ldots$$

(a) Draw and carefully label the state transition diagram.
(b) Find p_n as a function of p_0, λ, and μ.
(c) Develop a closed form expression for p_0.

34. Consider the cyclic queueing system of Figure 3.24.
(a) Draw the state transition diagram if there are $N = 3$ customers. Let the state variable be the number of customers in the upper queueing system. Redraw the state transition diagram if λ is 1.0 and μ is 2.0.
(b) Solve for p_1, p_2, and p_3 in terms of p_0. Solve for p_0 numerically.
(c) Find the mean throughput of the upper queueing system.

35. Consider a queueing-based memory model system as in section 3.3.4. Let there be two classes of customers. Class 1 customers have a mean arrival rate of 10 requests per second, a mean service rate of 20 requests serviced per second, and 1 Mbyte of memory is needed per request. Class 2 customers have a mean arrival rate of 5 requests per second, a mean service rate of 15 requests serviced per second, and 3 Mbyte of memory is needed per request. The system has 1000 Mbytes of total memory.
(a) Write an expression for the probability that the system is in state $\underline{n} = (n_1, n_2)$ where $\underline{n} \in \Omega$. That is, $\underline{n}$ is an allowable state.

(b) How do the size of the memory requests enter into the model and solutions?

36. Consider an M/D/1 queueing system. This is an M/G/1 system where the service time is deterministic (the same constant service time for all customers). Modify the Pollaczek–Khinchin mean value formula for this case.

(a) How does it differ from the formula for the expected number of customers for an M/M/1 queue $[E[n] = \rho/(1 - \rho)]$?

(b) Tabulate the expected number of customers for the M/M/1 and M/D/1 systems for $\rho = 0.1, 0.2, 0.4, 0.6, 0.8, 0.9$, and 0.99. Which system has the larger expected number of customers? How large is the difference as $\rho \to 1.0$?

37. Consider an open queueing network as in Figure 3.14 but with an extra feedback path from the output of Q2 to its input. The independent routing probabilities for customers departing Q2 are $r_{21} = 0.25$, $r_{22} = 0.5$, and $r_{2d} = 0.25$.

(a) Solve the traffic equations for the mean throughputs of Q1 and Q2 as a function of the mean arrival rate λ.

(b) Write out the product form solution for the equilibrium probability $p(n_1, n_2)$.

38. Consider a closed network of two queues, Q1 and Q2, in a loop where there is also a feedback path from the output of Q2 to its input. The independent routing probabilities for customers departing Q2 are $r_{21} = 0.3$ and $r_{22} = 0.7$.

Solve the traffic equations for the relative mean throughputs of Q1 and Q2. Let Q1 be the reference queue with mean throughput equal to 1.0.

39. Develop equation (3.167) for closed Markovian queueing networks from the associated global balance equation and the traffic equations.

40. For the network of problem number 38, let $\mu_1 = 2.0$ and $\mu_2 = 4.0$. Run the mean value analysis algorithm for $N = 1, 2$ and 3.

41. Consider a Gelenbe-style negative customer network as in Figure 3.16. However let Q2 receive positive (Λ) and negative (λ) external arrivals. Solve for q_1, q_2 and the product form solution for $p(\underline{n})$.

42. Find equations (3.250) and (3.251) from balance type equations for the voice/data integrated protocol example of section 3.8.

43. Consider two finite buffer queues in tandem (series). Let the first queue hold at most N customers, and the second queue hold at most one customer.

(a) Draw the state transition diagram. The arrival rate is δ, and the state dependent service rate of the first queue is λ_i, where i is the number of customers in the first queue. Also the service rate of the second queue is μ. Let the horizontal axis represent the number of customers in the first queue, and let the vertical axis represent the number of customers (0 or 1) in the second queue.

(b) Write recursive equations, as in section 3.8, for the network's equilibrium state probabilities. There are two possible sets of equations.

44. Draw and label a Petri net of the following situation. Sometimes a patron in a library can't find a book on the shelves. The front desk assigns two pages to look for the book, one on each of the two floors of the library. Assuming that the book is found by one of the pages, that page finds the other page and they both return to the front desk with the book.

45. Consider a "Dining Philosophers" stochastic Petri net. In this classic distributed system problem (Dijkstra 68) five philosophers are seated around a circular table. Between each philosopher on the table is placed a single chopstick. A philosopher needs two chopsticks to eat. If a philosopher picks up a chopstick from both sides of him/her, the philosophers on either side of him/her cannot eat.

(a) Draw and clearly label the Petri net of this situation. Each philosopher may be either thinking or dining, represented by places. With both chopsticks on the table on either side of a thinking ith philosopher, he/she picks them up at rate q^{i0}. The ith dining philosopher releases both chopsticks at rate q^{i1}. Each chopstick's availability is represented by its own place.

(b) Draw and label the state transition diagram.

46. Consider a Markovian Petri net of a *single* user submitting a job to a computer system. The user has three states: idle (0), job request (1), and job being processed (2). The user can only move from a job request to the job being processed if two independent resources, the memory and the CPU, are free (available). Once the job is processed, resources are released. The timing of all transitions is Markovian (negative exponential). The Petri net is safe (i.e., all places have at most one token).

(a) Draw and clearly label the Petri net.

(b) Draw the state transition diagram.

(c) Solve the state transition diagram for the three (equilibrium) state probabilities. Provide a closed form solution for the equilibrium probability that the processor is idle (not processing a job).

4

Fundamental Deterministic Algorithms

4.1 Introduction

In this chapter some basic deterministic algorithms used in networking are described. This discussion starts with a consideration of routing, a network layer function. Two shortest paths routing algorithms are presented. This is followed by an exposition of some different types of routing strategies. Second, protocol verification and model checking is examined. Finally error codes, both error detecting and error correcting, are studied.

4.2 Routing

4.2.1 Introduction

Some types of computer networks do not have a "routing problem" as there is a single path between nodes. Routing internal to a token ring or Ethernet are examples. However, in a wide area or metropolitan area network with multiple potential routes between each source and destination pair, there definitely is a routing problem.

For the purpose of routing, and for many other purposes, networks are usually represented as graphs. That is, the nodes in a graph will model packet switches, telephone switches, or computers. The edges of a graph represent links (either wired or wireless).

How many potential routes are there in a network graph between a source and destination? Let's do an illustrative example. Consider the rectangular graph of Figure 4.1. We wish to find the number of direct routes (without loops) from node A to node Z.

Consider the indicated path. It can be seen that it consists only of movements up (U) in the graph or to the right (R). Thus the path can be represented by the "word"

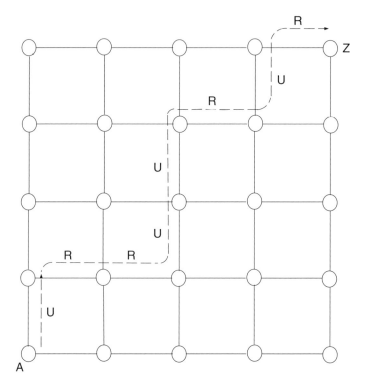

Fig. 4.1. A grid network with a specific shortest path from node A to node Z

$$URRUURUR \tag{4.1}$$

Notice that there are four U's and four R's. A little thought will show that any direct path from node A to node Z consists of eight letters with four U's and four R's in some pattern. Thus the number of possible paths is

$$\binom{8}{4} = 70 \ \text{ paths} \tag{4.2}$$

That is quite a large number of paths for such a small graph. We can generalize this. If we have a rectangular graph of $N \times N$ nodes, then by the same reasoning the number of direct paths is

$$\binom{2(N-1)}{N-1} \ \text{ paths} \tag{4.3}$$

We have in tabular form

Table 4.1. Number of Paths

N	No. of Paths
10	48,620
20	3.5×10^{10}
30	3.0×10^{16}
40	2.7×10^{22}

Table 4.2. Dijkstra's Algorithm

	N	B	C	D	E	F
1	{A}	12	∞	10	∞	2
2	{A,F}	12	3	10	∞	(2)
3	{A,C,F}	6	(3)	10	4	2
4	{A,C,E,F}	6	3	6	(4)	2
5	{A,C-F}	6	3	(6)	4	2
6	{A-F}	(6)	3	6	4	2

One can see from table 4.1 that the number of direct paths increases exponentially. It doesn't take a very big network to get a number of potential paths exceeding Avogadro's number!

If we measure the "cost" of a path by the distance along it, one can see that the "direct paths" of the previous examples are the shortest (distance wise) paths from node A to node Z. Such "shortest paths" are very desirable if one is routing packets or circuits.

However, distance is not the only way to measure cost. A link's "cost" may be in terms of quantities such as mean delay or monetary cost. Assume each link in a graph has a fixed cost. Then a "shortest path" between two nodes is a set of consecutive links connecting the two nodes such that the sum of costs of each of its links is the smallest possible over all possible routes between the two nodes. In fact there are relatively efficient algorithms for finding shortest paths. Two such algorithms are discussed in the next two sections.

4.2.2 Dijkstra's Algorithm

Consider the network of Figure 4.2. We wish to find all of the shortest paths from root node A to all of the other nodes. A table is created (Table 4.2) where one row will be added at a time. The set N in the table is the set of nodes for which we "know" a shortest path. The other column entries correspond to the current distance found for each node to node A.

The initial row has numerical entries for all nodes that are direct neighbors (one hop away) from node A. For Figure 4.2 This is nodes B, D, and F.

Dijkstra's genius here lies in proving (Tanenbaum 03) that the smallest entry in each row not selected yet corresponds to an optimal shortest path

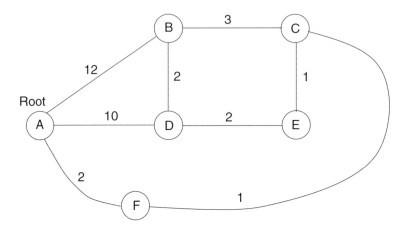

Fig. 4.2. A network graph with link costs indicated

distance. The smallest distance in row 1 is 2 for node F. In each row one attempts to improve the paths to the root node A for direct neighbors of the selected optimal node (indicated by parenthesis) of that row.

This can be written as a recursion (Schwartz 87)

$$D(v_j) = \min[D(v_j), D(w_i) + l(w_i, v_j)] \tag{4.4}$$

Here w_i is the node selected in the ith row and v_j are its direct neighbors. The current distance from node v_j to the root is $D(v_j)$. The distance from node w_i to v_j, over a single link, since they are direct neighbors, is $l(w_i, v_j)$. The equation says that in the current iteration (row) the new distance from node v_j to the root is the minimum of the old distance from node v_j to the root or the distance resulting from a route from node v_j to node w_i and from w_i back to the root.

So in our Table 4.2 node F is selected in row 2 and brought into set N. Since C is a direct neighbor of F with an entry of infinity, C's entry can be improved by going to the root from C to F to A with a total cost of 3. For the next row (row 3) node C's entry has the smallest value so it is selected. Nodes B and E are direct neighbors of C and their distances can be improved, so we can enter their new distances in the table to the root through node C (6 and 4, respectively). In the fourth row node E is selected and thus its direct neighbor D's entry improves from 10 to 6. In rows 5 and 6, nodes D and B are selected, respectively. However, no entries change in these two rows. It should be pointed out that in row 5 both D and B have entries of 6. It doesn't matter which one is selected first to be part of set N. Row 6 indicates the shortest path distances from each of the nodes to root node A.

Note that the Dijkstra algorithm procedure generates N rows in the algorithm table for an N node network. Note also that the algorithm naturally

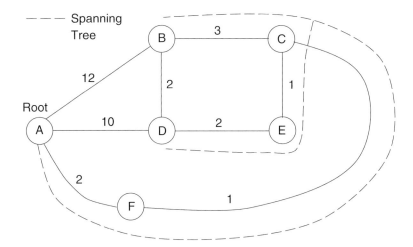

Fig. 4.3. A shortest paths spanning tree originating from the root (node A) super-imposed on the network graph of Figure 4.2

finds the shortest paths not just between a pair of nodes but from a root node to all of the other nodes.

If one needs the shortest distances between every pair of nodes, the algorithm is run N times, each time with a different root node. If the actual paths are desired, these may be carried along in the table in each entry as pointers to the next node along the current path back to the root (as is done in the next section). Also, it is possible for the links to be bidirectional with different costs in each direction (Saadawi 94). Finally, it should be noted that the routes found by the Dijkstra (or Ford Fulkerson) algorithm form a spanning tree. A spanning tree is a graph without loops that touches every node in the original graph. A shortest paths spanning tree for the network used in the previous example and its routing solution appears in Figure 4.3.

A use of the algorithm table is to generate a "routing table" that indicates which nodal output port leading to a direct neighbor to use to route packets/circuits to a distant destination. A routing table will be stored in each network node for routing purposes. The routing table for Figure 4.2's node A appears in Table 4.3.

In this particular example, all of the routes from node A to other nodes go through a single node, node F.

In some situations it may be desired to find the k shortest paths between nodes A and Z that do not share any links (i.e., are link disjoint). To do this the shortest path algorithm is run to find the shortest path. The shortest path is recorded, and its links are deleted from the graph. The shortest path algorithm is run again on the reduced graph to produce the second link disjoint shortest path, which is recorded and then removed from the graph. The

Table 4.3. "Routing" Table

Destination	Nearest Neighbor
B	F
C	F
D	F
E	F
F	F

Table 4.4. Ford Fulkerson Algorithm

	B	C	D	E	F
Initial	$(\cdot,\infty)$	$(\cdot,\infty)$	$(\cdot,\infty)$	$(\cdot,\infty)$	$(\cdot,\infty)$
1	(A,12)	(B,15)	(A,10)	(D,12)	(A,2)
2	(A,12)	(F,3)	(A.10)	(C,4)	(A,2)
3	(C,6)	(F,3)	(E,6)	(C,4)	(A,2)
4	(C,6)	(F,3)	(E,6)	(C,4)	(A,2)

process continues until all k link disjoint shortest paths are produced, if they exist.

4.2.3 Ford Fulkerson Algorithm

The Ford Fulkerson algorithm is a bit different from the Dijkstra algorithm in the details. However, its goal is much the same as the Dijkstra algorithm. The Ford Fulkerson algorithm will also find all of the shortest paths from a root node to each of the other nodes and create a spanning tree of routes in doing so.

In the Ford Fulkerson algorithm table (Table 4.4) for Figure 4.2, each entry has two parts. The first part is a pointer to the next node along the (current) path back to the root for that entry's node. The second part is the current distance along the path. The initialization row is filled with entries of $(\cdot, \infty)$ to indicate no path is selected yet. On a computer, infinity is just a number much larger than any likely entry.

The entries are filled in from left to right, top to bottom. For a node's entry, one attempts to improve the previous entry for that node by routing through the direct neighbor with the best total path back to the root.

The basic recursion for each algorithm table entry is (Schwartz 87)

$$D(v) = \min_{w_j}[D(w_j) + l(w_j, v)] \tag{4.5}$$

Here v is a node whose entry one is trying to compute and $D(v)$ is the distance along the path from node v to the root. Also, w_j is the jth neighbor of node v. Finally, $l(w_j, v)$ is the single link distance between direct neighbors

v and w_j. Thus the recursion says that the new entry of distance $D(v)$ for node v is the minimum over all of v's neighbors, the w_j's, of the previous distance for node w_j to the root plus the cost of getting from node v to w_j.

The initial entries one puts in the table are for direct neighbors of the root. In Figure 4.2, node B is a direct neighbor of the root so its entry in row 1 is (A,12). That is, go from node B to node A with a cost of 12. Each entry is based on the most recent information in the table (generally located to the left of the entry in the same row and to the right of the entry in the row above). So, in row 1, node C can use node B's existing entry and one has a route from C to B to A with a cost of 15. The process continues. As an example, for row 3's node D entry, the previous entry in the row above is (A,10). However, at this point in the table, there is an entry for node E in the row above of (C,4). Thus one can go from node D to E to C to F to A with a cost of 6 so the new entry for D is (E,6). If there is ever two equally good choices for the next node on the path back to the root, either one can be chosen.

One can see that the algorithm terminates when there are no changes in two consecutive rows. The number of rows in the completed algorithm table depends on the problem and will be smaller if columns are labeled left to right from nodes closest to the root to those furthest from the root.

Again, the shortest distances between every pair of nodes can be found by running the algorithm with a different root each of N times. Links may also be bidirectional.

This and the previous section describe bare bones shortest path routing algorithms. Implementing them in a distributed fashion on a dynamic network presents challenges. As routing tables are updated, packets may loop (e.g., travel in circular paths through the network). If the cost function in a packet switched network is mean delay, and routing costs are based on this, oscillations are possible. That is, lightly loaded links attract traffic and become heavily loaded, whereas traffic avoids heavily loaded links that then become lightly loaded. See Bertsekas 91 and Schwartz 87 for discussions of such problems and solutions for them.

At this point some routing options will be discussed (see Saadawi 94 for an alternative treatment).

4.2.4 Table Driven Routing

In table driven routing, information on routes is stored in tables at each node. The tables are updated using shortest path algorithms based on events (e.g., a link going down) or periodically (e.g., every X seconds or every Y events). Usually a combination of event based updating and periodic updating is used. Event based updating is sometimes referred to as inside updating, and periodic updating is sometimes referred to as outside updating.

In a packet switched network using a datagram mode of operation, a node reads an incoming packet's header for its destination address. The destination address is looked up in a "routing table" such as Table 4.3 (different from the algorithm tables of the previous two sections) to find which output port of the node the packet should be sent out over. The packet is then placed in the buffer for the output port. The routing table used here has been previously constructed from an algorithmic routing table. Note also that, if the output port speeds are B bps per output port, the nodal processor needs to be able to place packets in N output buffers at rate NB bps to keep all output links continually busy.

In a circuit switching based network, a circuit entering a node has an identification number. A table lookup based on the identification number allows the circuit to be continued (switched) out the appropriate output port. Virtual circuit based asynchronous transfer mode (ATM) packets also have an identification number for each virtual circuit that is carried in the packet header. A node determines this identification number and uses a table lookup to see to which output port a packet in the virtual circuit stream should be sent.

4.2.5 Source Routing

Nodes do not maintain routing tables under source routing. Rather, a source node will insert the route (i.e., nodes to be visited) for a packet into its header. Each node visited by the packet refers to this list to determine the next node to which to send the packet. The obvious question is how does the source know the path to insert into the packet header?

A centralized approach is for the network to have a path server (Saadawi 94). A path server monitors the network and computes shortest paths. A node wishing to send a packet contacts the path server for a path. Like any centralized scheme, two main drawbacks of this approach are that the path server is a single point of failure (i.e., if it goes down, the whole network is down) and that if the path server receives too many requests, it may be a performance bottleneck.

A distributed scheme is to use what are called path discovery packets. A source wishing to send a packet "floods" (See section below) the network with many path discovery packets that simply travel through the network without a specified path. Each node receiving a path discovery packet, before sending it to a neighboring node, puts its own node identification number at the end of a list of nodes visited by the packet in the packet's header.

The theory is that one or more of the flooded packets will eventually reach the intended destination node (which is also indicated in the path discovery packet's header). The destination node then has one or more routes to the source. A packet with either the first route received or some choice of the "best" route if several path discovery packets are received can then be sent back to source. The packet carrying the route can be source routed using the reversed list of nodes visited. Upon receiving the list, the original source node can launch a packet, or several packets, to the destination.

It should be pointed out that, if a discovered path is used too long by a source, network conditions may have changed and it may not be a good, or even feasible, route. On the other hand, too frequent path discovery burdens the network with flooding overhead.

4.2.6 Flooding

Flooding is a technique to get a packet(s) to a destination(s) without any, or very little, routing knowledge. In the simplest version of flooding, a node originating a flood sends copies of the packet(s) it wishes to send to all of its neighbors. A neighbor receiving such a packet(s) copies them out to all of its neighbors. That is, it will send copies of the (distinct) packets on all of its output ports except the one at which the packet(s) arrived.

Flooding may be used in situations where it is desired to broadcast a message to all of the nodes in a network (see the multicasting section below). It may be also used to get a message to a specific node, or a set of specific nodes, when there is no routing information available. It is also a good policy in a very unreliable network (where nodes and links go down frequently). However, the large number of packets generated is a large overhead, especially if just a small number of nodes needs to be contacted.

To reduce the number of packets generated, there have been strategies for flooding developed that flood only in limited directions (i.e., toward a destination). This is easier to implement if nodal geographic coordinates are known. This may be possible, even for mobile networks, if location systems such as a GPS (Global Positioning System) are used.

4.2.7 Hierarchical Routing

Hierarchical routing is a technique that allows routing table size reduction. It involves the way in which nodal addresses are assigned. Telephone numbers are an example of a hierarchical addressing scheme. In the United States, for instance, the first three digits of a ten-digit phone number comprise the geographic area code. The next three digits indicate the switching exchange, and the last four digits indicate the actual phone number within the indicated exchange.

The beauty of this system can be explained in terms of a simple example. Suppose that someone in San Francisco wishes to call a number in Manhattan, New York. The routing table used by long-distance facilities in San Francisco needs only store one entry for the millions of phones in the 212 Manhattan area code. Moreover, only a single entry is needed for the 10,000 phones in the destination local exchange. It should be mentioned that country codes add an extra level to the telephone hierarchy. The actual switching hierarchy of long-distance facilities and local exchanges is in fact a physical realization of the hierarchical addressing system.

Other than telephone networks, hierarchical routing can be done in other types of networks. One might have sets of wireless nodes grouped into "clusters" and clusters grouped into "super-clusters." The address 2.7.12 might indicate the twelfth node in the seventh cluster in the second super-cluster. There is a single entry in super-cluster routing tables for the second super-cluster nodes. Within clusters in the second super-cluster there is a single entry for the seventh cluster's nodes.

Hierarchical routing is very effective, particularly for large networks, at reducing routing table size. It also provides some intuitive structure to the address space, which is useful. On the downside, some hierarchical paths between nodes may be longer than direct connections, although this is less of a problem for large, dense, networks. There can also be a problem if a network under hierarchical routing grows with time (adds more nodes) and the space for the entries at each hierarchy level is limited — as in the current problem of proliferating telephone area codes in the United States.

4.2.8 Self-Routing

A number of networks with special structured topologies are known to have a useful property called "self-routing." As we shall see, self-routing means that packets can be routed from node to node in a network using only (part of) the destination address. Generally these self-routing networks or "switching fabrics" are implemented in very large-scale integration (VLSI) to serve as the routing heart of a packet switch. This form of switching is also known as space division switching as there are spatially separate paths through the switching network.

As an example consider the 8 input, 8 output (8×8) delta network of Figure 4.4. The inputs are on the left, and the outputs are labeled in binary on the right. The nodes are called switching elements. Each switching element has two inputs on the left and two outputs, labeled as "0" (upper output) and "1" (lower output), on the right. Thus, we have 2×2 switching elements. The fact that the same switching element circuitry can be replicated on a chip many times makes this a useful approach for VLSI implementation.

The wiring between switching elements is done in a patterned manner that allows self-routing. Specifically, a packet, say at the input, has the binary address for the intended output port for that packet placed in the packet's header. A switching element in the jth vertical column of switching elements receiving a packet will route it to the element's output with the same bit label (0 or 1) as the bit in the jth position (read left to right) of the destination binary address.

For instance, in Figure 4.4, a packet is launched from the third input from the top for output port 101. The switching element in the first vertical column of switching elements that the packet enters will route it to its lower (1) output, the switching element in the second column that the packet enters will route it to its upper (0) output, and the element in the last column

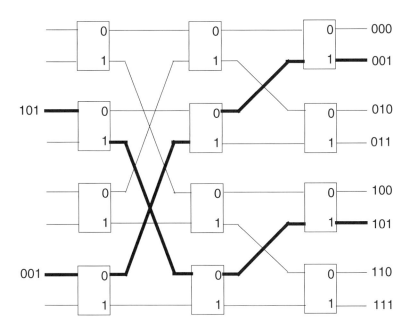

Fig. 4.4. An 8×8 delta network with two paths from specific inputs to outputs indicated

that the packet enters will route it to its lower (1) output and hence to the destination output port. A similar path is shown for a packet at the seventh input going to output 001.

Why does this work? Study the wiring pattern of the the 4×4 module in the second and third columns, third and fourth rows. This module is also a 4×4 self-routing delta network. For this module, it can be seen that the wiring is such that packets arriving at the 4×4 module inputs with a 0 in the middle bit are routed to the upper output (third stage) switching element. Packets with a 1 in the middle bit are routed to the lower output switching element of the 4×4 delta module. In both cases, the third bit then directs the packet to the proper output. This all matches the output port address pattern.

Looking again at the overall delta network, it can be seen that the upper and lower groups of four switching elements in the second and third columns consist of two (upper and lower) 4×4 delta networks. The wiring between the first and second stage is such that if the first bit in the output port address is a 0 the packet is sent from the first stage to the upper 4×4 delta network, no matter from which input it originates. If the first bit is a 1, an arriving packet to any first stage input is sent to the lower 4×4 delta network. Again, this all matches the output port address pattern. Larger switches can be recursively constructed in this manner.

It can be seen in this example that there is no need for routing tables or a centralized routing control. Routing decisions are made locally at each switching element based on a single destination port address bit in the packet header.

Such interconnection networks as the delta network were first studied in the context of telephone circuit switching and later applied to packet switching, particularly as ATM technology was developed (Robertazzi 93a). Delta networks are a subclass of the more general Banyan networks. Delta networks include omega, flip, cube, shuffle exchange, and baseline networks.

As Ahmadi and Denzel relate (Ahmadi 89), the major features of these networks are as follows:

• They have $\log_b N$ stages (columns) of switching elements and N/b switching elements for each stage. Here N is the number of inputs/outputs and b is 2.

• They are self-routing.

• They can be built in modular fashion using smaller switches. As was said, in Figure 4.4, in the second and third columns, the first and second (and third and fourth) row elements form 4×4 self-routing delta networks.

• They can be used in either synchronous or asynchronous mode.

• Their regular structure makes them attractive for VLSI implementation.

Although the $\log_b N$ complexity of the number of switch points seems better than the N^2 complexity of a crossbar switch (see chapter 2), the individual crosspoint circuit complexity for a crossbar is simpler so that in terms of chip area, Banyan and crossbars have been found in some studies to be comparable (Franklin 81, Szymanski 86).

With multiple packets being routed simultaneously through such self-routing interconnection networks, at times it will happen that the two packets entering a switching element's inputs have the same switching element output as their next destination. Although one packet may be accommodated at a time, the other will have to wait (be buffered). This phenomena is called blocking and networks that exhibit it are called blocking networks.

To mitigate the amount of blocking in a switching network and boost switch throughput, there are several approaches (Ahmadi 89):

• Use higher internal link speeds compared with external link speeds.

• Place buffers at each switching element.

• Use a handshake protocol between each stage and a "back pressure" strategy to slow down blocked packet movement.

• Use parallel networks so that one has a multiplicity of paths between each input and output. Alternatively, one can implement multiple links between each switching element.

• Preprocess the load through networks such as distribution or sorting networks. Distribution networks distribute load in a uniform manner to a switching network. Sorting networks sort packets by output port address. As part of a larger preprocessing system, sorting networks can be used to minimize blocking.

4.2.9 Multicasting

There are a number of ways in which packets can be multicasted (sent to multiple destinations) Packets can be flooded, although this is somewhat indiscriminate and incurs a large overhead. Alternatively, packets can be individually addressed to each destination. This is more efficient than flooding. Even more efficient than this at conserving network bandwidth is to put multiple addresses into individual packet headers. Assuming that nodes receiving such packets know the topology of the network, a multidestination packet arriving at a node is divided into a number of smaller multidestination packets, one for each nodal output port. The addresses in a divided packet are for nodes reachable by the output port the divided packet goes out over. The procedure repeats as packets arrive at nodes until copies are delivered to all individual destinations.

Finally, routers in networks can maintain a spanning tree(s) (see Figure 4.3) for Banyan network purposes. A special broadcast address allows packets to be forwarded at each node only to the spanning tree links.

Note that there are trade-offs in these techniques between the bandwidth consumed and the additional network control overhead needed to implement the more efficient schemes.

4.2.10   Ad Hoc Network Routing

Ad hoc networks are wireless networks of (usually) mobile nodes that hop packets from node to node along the way to their destinations. This is called multiple hop transmission. Energy conservation is an important part of ad hoc network design. In fact because of the nonlinear relation between transmission energy and radio propagation distance, it is more energy efficient for a packet to make several smaller hops rather than one large hop.

Routing algorithms for ad hoc networks can be divided into topology versus position based algorithms (Mauve 01).

Topology based algorithms can be further divided into proactive and reactive approaches. Proactive topology based routing algorithms use classic table based routing strategies. Information is continually maintained on paths that are available. A downside is that there is a large overhead in table update messages in maintaining the information for unused paths if there are frequent topology changes.

Reactive topology based algorithms only maintain routes that are currently in use. Naturally some sort of route discovery is necessary before a packet is transmitted. There may still be heavy update traffic with topology changes. Reactive protocols include DSR, TORA, and AODV (Murthy 04, Perkins 00, Tanenbaum 03, Toh 02).

Position based routing makes use of the geographic locations of nodes in making routing decisions. A location service such as GPS may be used. The position of a destination is placed in the packet header. A packet is forwarded closer and closer to the destination. Note that "geocasting" to a geographical region is straightforward. With position based routing there is no need for routing tables and their associated maintenance.

4.3 Protocol Verification

Protocols are the rules of operation of computer networks. A protocol specification is a set of rules for communicating between processes on different machines. Simple protocols may be expressed as state machines (see any text on digital logic) in state machine diagram form. However realistic protocols may have many states and transitions — too many to draw in a simple diagram. Thus an important question is whether a complex protocol has errors that may cause problems at some point in its operation such as a deadlock. The problem of checking a protocol specification for logic errors is called protocol verification or protocol validation. It is advantageous to catch such logic errors early in the design process, rather than once a system is implemented.

To make this discussion more concrete, consider the state machine representation of two communicating processes in Figure 4.5 (Yuang 88). A channel connects both processes in each of both directions.

Transmitted message identification numbers appear next to the transitions in the diagram. A negative ID number indicates a sent message, and a positive ID number indicates a received message.

Suppose that both processes start in state 1. Process 1 can send message 1, leading it to state 2. Process 2 can then receive message 1, leading it to its state 2. In a similar manner, message 3 can be sent and received (bringing both processes to their respective state 3's) and then message 4 can be sent and received (bringing both processes back to their state 1's). Alternatively, both processes can move from state 1 to state 2 and back to state 1 by sending and receiving packets 1 and 2 in sequence.

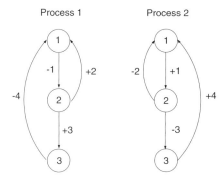

Fig. 4.5. State machine representation of a correct protocol. From Yuang © 1988 IEEE

The state machine of Figure 4.5 represents a "correct" protocol with no errors. However, a number of errors can arise in actual protocols. Among these are:

• *Unspecified Reception:* A message in the channel may be received but not as initially specified in the design. Therefore the system behavior at this point cannot be predicted accurately.

• *Deadlock:* The system is stuck or frozen in some state.

• *Livelock:* Messages are continually exchanged with no work accomplished.

• *State Ambiguity:* A state in a process can stably coexist (i.e., is reachable with empty channels) with several states in another process. This is not necessarily an error, but one must be careful with these situations.

• *Overflow:* The number of messages in a channel or buffer grows in an unbounded fashion.

• *Non-executable Interaction:* A transmission or reception that is indicated in the specification but is never executed in reality. Also called dead code.

Let's consider an example of the specification of two communicating processes that has some errors (from Yuang 88). The state machine diagram is shown in Figure 4.6. A "reachabilty diagram" that illustrates all possible states is shown in Figure 4.7. Each rectangular entry is a state. The entries in (row,column) positions (1,1) and (2,2) indicate the state of process 1 and process 2, respectively. The entry (1,2) indicates which message is in the

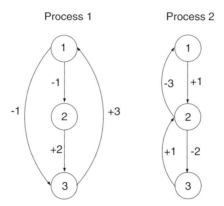

Fig. 4.6. State machine representation of a protocol with errors. From Yuang © 1988 IEEE

channel from process 1 to process 2. Likewise, the entry (2,1) indicates which message is in the channel from process 2 to process 1. An "E" indicates an empty channel.

The diagram begins in global state GS 1. Both processes are in their respective state 1's, and both channel directions are empty. Consider the left branch of the reachability diagram. Message 1 is put on the channel from process 1 to process 2, and process 1 moves into state 2 (GS 2). This message is received by process 2, leading to both processes being in their respective state 2's and both channel directions are empty (GS 3). Next, message 2 is put on the channel from process 2 to process 1 (GS 4) and process 1 receives it (GS 5). Now both processes are in their respective state 3's and both channels are empty (GS 5).

But there is now a problem. Referring to the state machine diagram (Figure 4.6), process 1 expects to receive message 3 and process 2 expects to receive message 1, neither of which will be sent. This is a deadlock! The system is permanently stuck in this state.

Now consider the right branch of the reachability diagram. From GS 1, message 1 is put on the channel from process 1 to process 2 and process 1 moves into state 3 (GS 6). Process 2 receives the message (GS 7). With the system in GS 7, process 1 is in its state 3, process 2 is in its state 2, and both channel directions are empty. Two things can happen now. Process 2 can launch message 3 to process 1 (GS 8) followed by its reception by process 1, leading to the system returning to original state GS 1 (which is perfectly fine). Alternatively, from GS 7 process 2 can launch message 2 to process 1, leading to GS 10.

In global state GS 10, both processes are in their respective state 3's and message 2 is on the channel from process 2 to process 1. Again we have a deadlock. Process 1 is expecting message 3, and process 2 is expecting message 1,

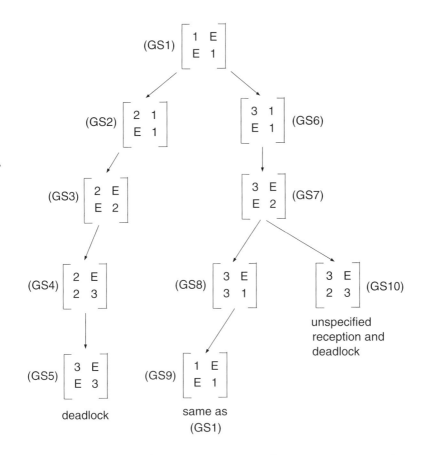

Fig. 4.7. Reachability tree of state machine protocol in Figure 4.6. From Yuang © 1988 IEEE

neither of which will be sent. Moreover as there is a message on the channel to process 1 that process 1 is not prepared to receive, we also have an unspecified reception. Note also that the transition from state 3 to state 2 in process 2 is never executed. It is a non-executable interaction (dead code).

Through this example one can appreciate that, if the reachability diagram has hundreds or thousands of states, which can be true of even a system of moderate complexity, finding deadlocks and other errors is a challenging algorithmic problem. Speaking generically, one can implement:

- *Exhaustive Search:* Search the entire state space, although this is impractical for larger protocols.

- *Local Search:* Search local parts of the state space.

- *Probabilistic Search*: Search states with a high probability of occuring.

- *Divide and Conquer:* Break the problem into smaller parts.

As an example, the SPIN model checker software has been highly optimized over the years to be very efficient at finding protocol errors (Holzmann 97,04). It should be pointed out that an algorithm can either be designed to prove that there are no errors or to find errors. Most algorithms find errors as it is more tractable.

4.4 Error Codes

4.4.1 Introduction

The normal movement of electrons at any temperature causes thermal noise in electrical circuits. Lightning strikes cause impulses of noise (impulse noise) that interfere with radio transmission. Two wires that are physically close may be electromagnetically coupled, causing the signals in each to mix and thus cause cross-talk interference. Finally, there is optical noise in fiber-optic cables due to the indivisible nature of photons.

When a binary stream of data is transmitted, such mechanisms in the channel may cause 1's to become 0's or 0's to become 1's. Recall that a 0 or 1 will be represented by a distinct waveform. Electrical, radio, or optical noise, which distorts such a waveform by adding a random-like component, may cause the receiver to make a mistake in waveform recognition, thus allowing a "0" to be received as a "1" or vice versa.

How does one then reliably transmit data in the presence of noise? How can one be sure that a transmission of financial data is accurate, for instance? The solution to this problem that people have come up with over the years is the use of error codes. Using a mathematical algorithm at the transmitter, extra check bits are added to each block of data (packet). An inverse version of the algorithm is run by the receiver on the received data and check bits.

There are two types of error codes. An error detecting code allows the presence of (certain types of) errors to be detected at the receiver. Under an error detecting code, the receiver will not know which bits in the block have errors, only that there are some. In this case the receiver usually asks the transmitter to retransmit the data block that originally had errors. This protocol is called Automatic Repeat Request (ARQ). A more powerful error correcting code allows the receiver to correct (certain types of) errors on the spot without asking for a retransmission. This is particularly efficient when propagation delays are large. As an extreme example it would be very inefficient for a space probe to Saturn to ask for a retransmission if an error(s) were detected when the round-trip delay is measured in hours.

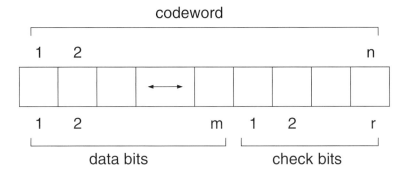

Fig. 4.8. A code word with message and check bits

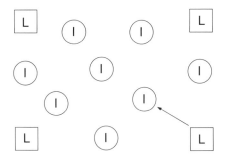

Fig. 4.9. A legitimate code word (L) transformed by an error(s) into an illegitimate (I) code word for an error detecting code

Consider m bits of data with r check bits for a total of $m + r = n$ bits as in Figure 4.8.

A basic concept involving the difference between two n-bit code words is Hamming distance. The Hamming distance between two code words of equal size is the number of bit positions that are different. For instance, the code words 1011 and 1101 are a distance 2 apart as two bit positions differ.

In Figure 4.8, all 2^m data possibilities may be used but not all 2^n code word possibilities are used. In an error detecting code, some combinations of data and check bits are legitimate and some are not. Consider a "space" of all possible code words as in Figure 4.9. As illustrated in the figure, an error will hopefully change a legitimate code word in the space (indicated by a boxed L) to an illegitimate one (indicated by a circled I), which can be recognized as such by the receiver.

Generally a code will protect against some degree of the most likely errors. As an example, a parity code can protect against single bit errors. A cyclic

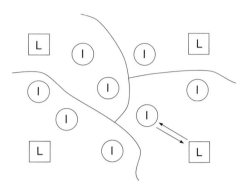

Fig. 4.10. Space of legitimate (L) and illegitimate code words (I). A legitimate code word is mapped by an error into an illegitimate code word. The illegitimate code word is mapped to the nearest (and correct) legitimate code word

redundancy check (CRC) code detects single, double, and odd numbers of bit errors as well as certain burst errors. It is always possible for an error code to be overwhelmed by too many errors. For instance, in Figure 4.9, if an error is such that it causes a legitimate code word to be transformed into another legitimate code word, this will escape detection at the receiver.

A diagram similar to figure 4.9 for error correcting codes appears in Figure 4.10. In this figure, each legitimate code word in the code space is surrounded by a neighborhood of illegitimate code words. When the receiver detects an illegitimate code word it assumes the closest legitimate code word is the correct one. Closeness here may be in a Hamming distance sense. Naturally if a legitimate code word, is distorted too much it may become a code word in a different neighborhood and be mapped into a wrong but legitimate code word.

We now discuss three block codes in detail. Note that, although it is beyond the scope of this book, stream coding is also possible.

4.4.2 Parity Codes

This is a simple code that can detect single bit errors. One adds a single check bit to the data block such that the number of 1's in the entire code word (data bits plus check bit) is even (if one wants to use "even parity") or odd (if one wants to use "odd parity"). We will always use even parity in examples in this chapter.

For instance, with even parity, 1011 becomes 10111 and 00101 becomes 001010 where the last bit is the appended parity/check bit. The receiver simply counts the number of 1's in the received block. If it is even, the message is assumed to be correct; if it is odd there has been an error. In the case of an error a request to the transmitter for a retransmission is usually sent by the receiver. Parity codes will be used as an element of the Hamming error correcting codes of the next section.

A burst error is a series of errors affecting a number of consecutive bits. For instance, a lightning strike of a certain duration may cause a burst error in a serial transmission. A trick can be used to still use parity codes in the presence of burst errors.

For this coding trick, arrange a number of code words in a table with one code word per row. But transmit a column, not a row, at a time over the serial channel. Then if a burst error occurs in the channel that is not too long in duration, only one bit in a number of consecutive code words will be affected and the errored code words will be detected.

It should be noted that errors can occur in both data and check bits. Therefore any code must be able to handle both types of errors. It can be seen that this is true of parity coding, for instance.

4.4.3 Hamming Error Correction

In this section we'll take a detailed look at a 1-bit Hamming error correcting code (Tanenbaum 03). This code can correct 1-bit errors at the receiver with 100% accuracy. The code uses parity bits as building blocks.

Before proceeding to the actual coding mechanism, a question that needs to be answered is how many check bits are needed for a given number of data bits. Recall there are m data bits, r check bits, and $n = m + r$ total bits in a code word. There are 2^m legal messages (i.e., all possible messages in m bits). Each such legal message is associated with itself and n possible corrupted messages that are corrupted by one bit being flipped to its opposite value. Thus 2^m times $(n + 1)$ should be less than the total number of code words (2^n) or

$$2^m(n + 1) \leq 2^n \qquad (4.6)$$

Let $n = m + r$ and

$$2^m(m + r + 1) \leq 2^{m+r} \qquad (4.7)$$

Divide both sides by 2^m, one obtains

$$(m + r + 1) \leq 2^r \qquad (4.8)$$

This equation needs to be satisfied for a given number of message bits m by the number of check bits r. For instance, suppose that $m = 10$ (ten message bits), and it is desired to find r. We can start with $r = 1$ and keep incrementing it until the inequality holds.

Thus four check bits are needed for ten data bits. As the number of data bits increases, the percentage overhead in check bits decreases.

How does a transmitter implement Hamming error code correction? We'll set up the code word so that check bits appear in bit positions that are powers

Table 4.5. Computing Number of Checkbits

r	Inequality	Holds?
1	$12 \leq 2$	No
2	$13 \leq 4$	No
3	$14 \leq 8$	No
4	$15 \leq 16$	Yes!

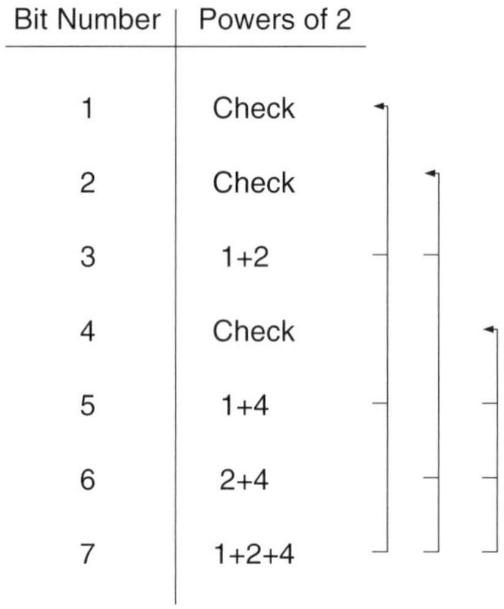

Bit Number	Powers of 2
1	Check
2	Check
3	1+2
4	Check
5	1+4
6	2+4
7	1+2+4

Fig. 4.11. Coverage of parity bits in Hamming code in a 7-bit code word

of two (i.e., 1, 2, 4, etc. . .). For four data bits and the required three check bits, this is illustrated in Figure 4.12.

In the Hamming code a number of parity bits, each covering overlapping parts of the code word, are placed in the powers of two bit positions. To see which check/parity bits are associated with which code word bits, consider the table of Figure 4.11. The seven bit positions are listed in the first column. Bit positions 1, 2, and 4 are listed as check bits. Each bit position number is expanded as a sum of numbers that are powers of two. For instance, 5 is $1 + 4$ and 7 is $1 + 2 + 4$.

Now each check bit will be (we'll say) an even parity bit for itself and bits where the check bit's position number appears in the sum of powers of two. Thus, for instance, check bit 1 covers bits 1, 3, 5, and 7. The check bit in bit position 2 covers bits 2, 3, 6, and 7. Finally the check bit in bit position 4 covers bits 4, 5, 6, and 7.

A transmitter implementing the 1-bit Hamming error correction code will format the code word and insert the check bits (although check bits are often grouped together at the end of a packet after the data bits for transmission).

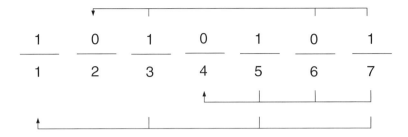

Fig. 4.12. Four-bit message (1101) embedded in a Hamming code word with check bit coverage shown

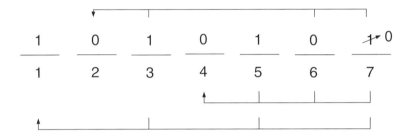

Fig. 4.13. Seven-bit Hamming code word with bit in seventh position flipped by noise from 1 to 0

As an example, consider Figure 4.12. The four message bits (1101) need three check bits for a total of 7 bits in the code word. From the figure one can see that the first check bit should be a 1 under even parity and the number of 1's then in bit positions 1, 3, 5, and 7 is even. Likewise the bit position 2 check bit is 0 so that the number of 1's in bit positions 2, 3, 6, and 7 is even. Finally, the check bit in position 4 is set to 0 so that the number of 1's in bit positions 4, 5, 6, and 7 is even. These seven bits (1010101) are then transmitted by the transmitter.

Let's see how a receiver can correct a single bit error. Instead of receiving the correct code word, 1010101, say a bit is flipped by channel noise so 1010100 is actually received (Figure 4.13).

The receiver checks whether the check bits and the bits they cover still have even parity. For this received code word each group of four bits associated with a check bit has odd parity. Since check bits 1, 2, and 4 are now associated with (incorrect) odd parity, one adds $1 + 2 + 4 = 7$ so that the receiver knows that bit 7 is in error. Thus the receiver will change the incorrect 0 in bit position 7 to a 1, and so error correction is achieved.

One can see the receiver procedure. For instance, if bits 1 and 2 only are associated with odd parity, the error is in bit position $1 + 2 = 3$ or bit 3. If only bit 4 is associated with odd parity, only the check bit in position 4 has an error.

Why does the Hamming code procedure work? By way of an example from Figure 4.11, if bit 5 has an error, checks bits 1 and 4 will be made odd so adding the odd bit position numbers $(1 + 4 = 5)$ will indicate that bit 5 has an error. This method will work for a single bit error in any bit position.

Since check bits only appear in powers of two bit positions, the number of check bits becomes a smaller proportion of the code word size as the code word length increases. For instance, in a 1023-bit code word there are only ten check bits.

4.4.4 The CRC Code

Introduction

Error detecting codes allow a receiver to detect the presence of an error(s), although the receiver will not know which bit(s) are in error and so can not correct the error. However the receiver can ask for a retransmission that hopefully will be error free (otherwise the receiver will ask for a second retransmission and so on...).

Error detecting codes generally require less overhead in the form of the number of check bits than error correcting codes. This can be true even if the overhead of retransmissions is accounted for.

The CRC code (Tanenbaum 03) is a powerful error detecting code that can detect single bit errors, double bit errors, odd numbers of errors, and many types of burst errors. It is used in Ethernet.

The CRC Algorithm

Cyclic redundancy codes are based on polynomial arithmetic. This sounds complicated but really is not. For instance, the binary number 1100110 can be represented as

$$1x^6 + 1x^5 + 0x^4 + 0x^3 + 1x^2 + 1x^1 + 0x^0 \tag{4.9}$$

Here each bit is a coefficient of a power term in x.

In a CRC code, both the transmitter and the receiver agree to use some "generator polynomial" $G(x)$ as the basis of the error detection.

The division of polynomials is used in CRC codes. In short, the transmitter selects check bits to append to the message (data bits) such that the resulting code word polynomial is exactly divisible by $G(x)$ (i.e., there is a remainder of zero). The receiver divides the polynomial of its received code word by the same $G(x)$. If there is a zero remainder, the receiver assumes that there is no error. If the receiver's division does produce a remainder, the receiver assumes that there must be an error.

		A B A±B

A B A±B
0 0 | 0
0 1 | 1
1 0 | 1
1 1 | 0

```
 1111
-1011
─────
 0100
```

Fig. 4.14. Exclusive or (XOR) truth table and 4-bit exclusive arithmetic example

Even though the CRC code is based on binary numbers, a base 10 example can make this idea more concrete. Suppose that the message is 27 and the generator number is 25. The transmitter appends a check digit to 27 to create 270. It then adds 5 to the number so that the result, 275, is exactly divisible by the generator number. Then 275 is transmitted. The receiver divides the received number by 25. If there is a remainder, it is assumed that there is an error. If there is no error, it is assumed that correct reception has been achieved. In this case the message 27 is recovered from the code word.

Recall that in doing division one needs to periodically subtract numbers. For CRC codes this is done in a special way using exclusive or arithmetic with no carries. In Figure 4.14 the exclusive truth table is illustrated. One can see that 0-0 is 0, 0-1 is 1, 1-0 is 1, and 1-1 is 0. In the same figure next to the truth table is a 4-bit example. The numbers in each column are subtracted independently of what happens in other columns (there are no carries). Thus $1111 - 1011 = 0100$.

Now for the detailed procedure. Let $M(x)$ be the message (data) polynomial $G(x)$ be the generator polynomial and $T(x)$ be the polynomial of the transmitted code word. We

(**a**) Append r zero bits to $M(x)$ to create $x^r M(x)$. Here r is the number of bits associated with the generator polynomial minus one. The number of check bits will also be r.

(**b**) Divide $G(x)$ into $x^r M(x)$.

(**c**) Subtract the remainder from $x^r M(x)$ to create $T(x)$.

The last step is necessary so that $T(x)$ is exactly divisible by $G(x)$. An example of a CRC code transmitter division appears in Figure 4.15. The generator is 10011 and the message is 10000. Four check bits are added (the number of generator bits minus one). The result of the division with its exclusive or subtractions is a remainder of 0101. This remainder is subtracted from $x^r M(x)$ (see bottom of Figure 4.15). But note that the result of subtracting a remainder from the (zero) check bits is just the remainder. Thinking about this and the exclusive or truth table, one can see that this always true. Thus the receiver doesn't actually do the subtraction; it simply substitutes the remainder into the check bit positions.

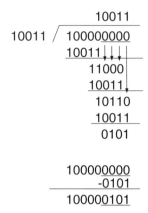

Fig. 4.15. CRC code transmitter example

What of the receiver? It simply divides $G(x)$ into what it receives $R(x)$. It does not add check bits; these are already incorporated into $R(x)$. If there is a remainder, the receiver assumes that the code word has an error and requests a retransmission. If there is a zero remainder, the receiver assumes that there is no error.

In practice CRC code division is most efficiently implemented using shift register hardware (Peterson 61, Tanenbaum 03).

CRC Code Protection

The CRC code protects against single bit errors, double bit errors, odd numbers of errors, and many types of burst errors. In this subsection it is demonstrated why this is true.

The polynomial associated with the transmitted code word is $T(x)$. We'll define an error polynomial $E(x)$ that has a 1 in every bit position where there is an error. For instance, with $E(x) = 100010$, there are errors in the first and fifth positions. However, $E(x)$ is an analytical tool. The receiver does not know which bit position is in error. What is received is $R(x) = T(x) + E(x)$ where exclusive or addition is used. From the exclusive or truth table (Figure 4.14), one can see that, if a bit in $E(x)$ is a zero (no error), adding it to the corresponding $T(x)$ bit simply leaves the $T(x)$ bit unchanged. On the other hand, if a bit in $E(x)$ is a 1 (error), the corresponding bit in $T(x)$ is inverted (1 to 0 or 0 to 1, an error).

Since $T(x)$, the code-word the transmitter transmits, is exactly divisible by the generator polynomial $G(x)$ one has at the receiver when $R(x)$ is divided by $G(x)$

$$\frac{R(x)}{G(x)} = \frac{T(x) + E(x)}{G(x)} \rightarrow \frac{E(x)}{G(x)} \qquad (4.10)$$

We now examine each type of error that the CRC code can detect. This basically comes down to the question of selecting $G(x)$'s that give the most protection against errors. which involves a consideration of whether $E(x)/G(x)$ has a remainder. See Tanenbaum 03 for an alternative treatment.

Single Bit Errors: For a single bit error there is a single 1 in some position (say the ith position). Now $E(x)$ is a power of two ($E(x) = x^i$). If $G(x)$ has two or more nonzero terms, it won't divide $E(x)$ evenly so a remainder will be produced and the error will be detected. In base 10 terms, if $E(x)$ is 32 (a power of two) and $G(x)$ is 7 (1+2+4, three terms), then 32/7 produces a remainder. Thus any good $G(x)$ should include two or more nonzero terms so that the receiver can detect single bit errors.

Double Bit Errors: In this case $E(x)$ has two nonzero terms

$$E(x) = x^i + x^j \tag{4.11}$$

$$E(x) = x^j(x^{i-j} + 1) \tag{4.12}$$

$$E(x) = x^j(x^k + 1) \tag{4.13}$$

Now consider $E(x)/G(x)$. One should select $G(x)$ so (a) it does not divide x^j evenly (has two or more terms, see the previous subsection) and (b) $G(x)$ does not divide $x^k + 1$ evenly up to some maximum k. Since $k = i - j$, this is the maximum distance between two bit errors (i.e., maximum packet length) that can be tolerated. Small polynomials that can protect packets thousands of bits long are known by mathematicians.

Odd Number of Errors: If there is an odd number of errors, $E(x)$ will have an odd number of nonzero terms. We'll use the observation that there does not exist a polynomial with an odd number of terms that is divisible by $(x + 1)$. So to detect an odd number of errors, all one has to do is be sure $(x + 1)$ is factor of the $G(x)$ used.

The observation, which is not obvious, can be proven by contradiction (Tanenbaum 03). Assume that $E(x)$ has an odd number of terms and *is* divisible by $(x + 1)$. Then certainly $(x + 1)$ can be factored out

$$E(x) = (x + 1)F(x) \tag{4.14}$$

Now let $x = 1$

$$E(1) = (1 + 1)F(1) \tag{4.15}$$

But with exclusive or arithmetic $1 + 1 = 0$ so

$$E(1) = (1 + 1)F(1) = 0 \times F(1) = 0 \tag{4.16}$$

However a little practice with simple examples will show that, if $E(x)$ has an odd number of terms, $E(1)$ always equals 1 (for instance, $E(1) = 1+1+1 =$

1). Again, we are using exclusive or arithmetic. Thus we have a contradiction in our original assumption, and the observation that a polynomial with an odd number of terms is not divisible by $(x + 1)$ is correct.

So any good CRC generator polynomial should have $(x + 1)$ as a factor to catch an odd number of errors.

Burst Errors: A burst error, as said earlier, is an error consisting of a number of consecutive errored bits. In this case the code word associated with $E(x)$ might look something like this:

$$0000000111100000000 \tag{4.17}$$

or

$$E(x) = x^i(x^{k-i} + x^{k-i-1} + \cdots + x^1 + 1) \tag{4.18}$$

Here x^i is a shift of the burst. In selecting $G(x)$ one can use the fact that, if $G(x)$ contains a term of $x^0 = 1$, then x^i is not a factor of $G(x)$ so that, if the degree (i.e., highest power) of the expression in parenthesis is less than the degree of $G(x)$, there will be a remainder.

Burst errors of length less than or equal to r will be detected if there are r check bits. Thus all one needs to do to accomplish this is to include a term of 1 in $G(x)$.

There is also some protection against larger bursts. For instance, if the burst length is greater than $r + 1$ or if there are multiple, smaller bursts, then the probability that the receiver doesn't catch an error is $(1/2)^r$. The same probability is $(1/2)^{r-1}$ if the burst length is $r + 1$. See Tanenbaum 03 for more details.

Certain generator polynomials are international standards. They all have more than two terms, they all have $(x + 1)$ as a factor, and they all have a term of $+1$. Among them are

$$x^{12} + x^{11} + x^3 + x^2 + x^1 + 1 \tag{4.19}$$

$$x^{16} + x^{12} + x^5 + 1 \tag{4.20}$$

4.5 Conclusion

In this chapter we have looked at fundamental deterministic algorithms. Keep in mind, though, that they are usually used in very stochastic (random) environments. That is, traffic flow and routing patterns, which states a protocol executes and channel errors can all be very well modeled as being random-like.

4.6 Problems

1. How many shortest paths are there in a rectangular mesh network of size 5 × 10 nodes between opposite diagonal corner nodes? Show the calculation.
2. Does a shortest path algorithm like the Dijkstra or Ford Fulkerson, find the shortest path between two nodes, or is it more general?
3. How can one find the shortest paths between all pairs of nodes using the shortest path algorithm?
4. How does one use the shortest path algorithm to find the k shortest link disjoint paths between two nodes?
5. What is the difference between the "algorithm" and the "routing" tables in this chapter? How are these used in table driven routing?
6. Make a "routing" table for node C in Figure 4.3.
7. How are paths found in source routing?
8. What is flooding? How can the number of packets sent in flooding be reduced?
9. Why are paths between nodes in hierarchical networks sometimes longer than direct connections?
10. Why is the concept of switching elements useful for VLSI design?
11. What is a disadvantage, in terms of implementation, of some of the multi-casting techniques mentioned in the chapter (specifically putting multiple addresses into packets and the use of spanning trees)?
12. Explain why it is more energy efficient for a packet to make several smaller hops rather than one large hop in an ad hoc network.
13. Under what condition(s) are reactive routing algorithms more efficient than proactive routing algorithms?
14. Why is protocol verification a challenging problem for large networks?
15. What is the difference between a deadlock and a livelock?
16. What is an unspecified reception?
17. What is the Hamming distance between 00001111 and 11001100?
18. In Figure 4.10 can a legitimate code word be mapped into another legitimate code word through some error? What is the result?
19. Can a parity code detect an odd number of errors?
20. What happens in the base 10 example of "The CRC Algorithm" subsection of section 4.4.4 if the code word 250 is received?
21. Run both the Dijkstra and the Ford-Fulkerson algorithms to create the algorithm tables for the network of Figure 4.16. Here node A is the root. Label the columns in each table, from left to right, as B, C, D, E, F, and G.
22. Make a copy of Figure 4.4 and indicate the paths taken by packets entering inputs 1 and 6, destined for output 011.
23. Create the reachability tree for the state machine protocol representation of Figure 4.17. Indicate any states involved in a deadlock or unspecified reception.

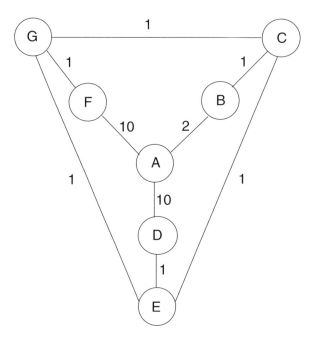

Fig. 4.16. A routing problem

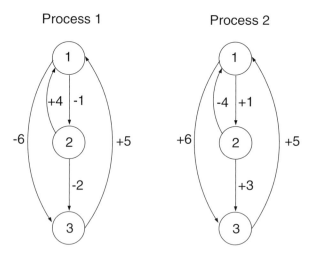

Fig. 4.17. A reachability tree problem

24. A parity bit is appended to 6 bits. Even parity is used. If the received code word is 1011000, is there an error?
25. A block of data is transmitted using a Hamming code. The data are 1011. Find the complete code word sent by the transmitter. Use even parity.

26. Four bits of information are received using a Hamming code. The received code word is 0010010. Assume the use of even parity. Is there an error? If so, in which bit is the error located? Check bits are in powers of two positions.

27. Consider a CRC code transmitter problem. Find the checksum to append to the message bits 1110001. Use the generator code word 10011.

28. Consider the received code word 100010111. Is there an error? There are four check bits at the end of the code word. The generator code word is 10011.

5

Divisible Load Modeling for Grids

5.1 Introduction

The increasing prevalence of multiple processor systems and data intensive computing creates a need for efficient scheduling of computing loads and related transmissions. An important class of such loads are divisible ones: data parallel loads that are completely partionable among processors and links.

Over the past decade or more a new mathematical tool has been created to allow tractable performance analysis of systems incorporating both communication and computation issues, as in parallel and distributed processing (Bharadwaj 96b, Bharadwaj 03, Robertazzi 03). A key feature of this divisible load scheduling theory (known as "DLT") is that it uses a linear mathematical model. Thus, as in other linear models, such as a Markovian queueing theory of electric resistive circuit theory, divisible load scheduling theory is rich in such features as easy computation, a schematic language, equivalent network element modeling, and numerous applications.

In divisible load scheduling theory, it is assumed that computation and communication loads can be partitioned arbitrarily among a number of processors and links, respectively. In addition, there are no precedence relations among the data. Therefore load can be arbitrarily assigned to links and processors in a network. Thus the theory is well suited for modeling a large class of data intensive computational problems, such as in grid computing. It also has applications for modeling and scheduling for some meta-computing applications and parallel and distributed computing problems. As a secondary benefit, it sheds light on architectural issues related to parallel and distributed computing. Moreover, the theory of divisible load scheduling is fundamentally deterministic. Although stochastic features can be incorporated, the basic model has no statistical assumptions that can be the Achilles' heel of a performance evaluation model.

This section of this chapter describes reasons why divisible load scheduling theory is a tractable, flexible, and realistic modeling tool for a wide variety of applications. Moreover, a tutorial introduction to modeling using divisible load theory follows. It includes a discussion of single level tree modeling, equivalent processors, the performance of infinite size networks, time-varying modeling, and the linear programming solution of divisible load scheduling models.

Since the original work on this subject in 1988 (Agrawal 88, Cheng 88), a fairly large number of journal papers and two books containing an expanding body of work on this subject have been published on a worldwide basis. The original motivation for some of this work was "intelligent" sensor networks doing measurements, communications, and computation. However most recently mentioned applications involve parallel and distributed computing.

A typical divisible scheduling application might consist of a credit card company that each month needs to process 10 million accounts. It could conceivably send 100,000 of the records to each of 100 processors. Note that simply splitting the load equally among the processors does not take different computer and communication link speeds, the scheduling policy, the interconnection network, or other features into account and so is a suboptimal policy in terms of solution time (and speedup). Divisible load scheduling theory provides the mathematical machinery to do solution time optimal processing.

Some other typical examples follow:

(1) A bank, insurance company, or online service may want to process large numbers of customer records for such purposes as billing, data mining, targeted direct mail advertising, or to evaluate the profitability of new policies/services. A mid-size cap fund would have a need to process complex financial records of many companies in order to make the best investment decisions or evaluate new investment strategies.

(2) An individual photograph typically needs to be represented by a great deal of digital information. Satellite imagery in particular can generate many thousands of such images in relatively short amounts of time. It is physically/economically impossible for humans to look at all such images in their entirety and in detail. One may want to process such images for particular patterns or features for such purposes as oil/gas exploration, weather, or planetary exploration. Another example of divisible image processing is searching millions of fingerprints or facial recognition records for a match.

(3) Engineers and scientists working at corporations, research labs, and universities have a need to process large amounts of data for various reasons (engineering studies, looking for particular patterns, public health studies). In fact large physics collider experiments at government-funded research labs are on the verge of generating petabytes (thousands of gigabytes) of data a year, all of which must be processed. Given increasing sensor and data collection capabilities, even modest engineering experiments can generate copious amounts of data.

(4) With advances in sensor design and implementation and the increasing prevalence of multiple processor systems (in everything from cars to scientific

equipment) there is a need for processing and performance prediction of sensor generated loads.

Thus there are many potential situations where a tractable and accurate approach to load scheduling would be useful. Ten advantages of using divisible load scheduling theory for this purpose appear below (also see Robertazzi 03).

TEN REASONS

Ten salient reasons to consider using divisible load theory will now be discussed.

(1) A Tractable Model: The key to doing optimal divisible load scheduling is what is called the optimality principle (Cheng 88, Bharadwaj 96b). That is, intuitively, if one sets up a continuous variable model of scheduling and assumes that all of the processors stop computing at the same time instant, one can solve for the optimal amount of total load to assign to each processor/link using a set of linear equations or in many cases, as in queueing theory, recursive equations. The model accounts for heterogeneous computer and link speeds, interconnection topology, and scheduling policy. Moreover the model may include fixed delays, such as a propagation delay in links. The model also can simply handle loads with different computation and communication intensities. Gantt chart-like schematics easily display a schedule being considered.

The relation among queueing theory, electric circuit theory, and divisible load theory is particularly interesting. In their basic form all three theories are linear ones. There are a number of commonalities, which include a schematic language, recursive or linear equation solution, the concept of equivalent networks, the possibility for time-varying modeling, and the possibility of results for infinite-sized systems.

The tractable nature of divisible load theory is in contrast to the nature of the more traditional indivisible load problem. That is, if one has to assign atomic jobs/tasks, each of which must run on a single processor, then one has a problem in combinatorial optimization that often is non-polynomial time (NP) complete (Bokhari, Murthy, Shirazi). Precedence relations provide an additional complication. It should be emphasized that divisible load theory is not applicable to all computer scheduling problems, but it is applicable to an important class of such problems.

(2) Interconnection Topologies: Over the years divisible load modeling has been successfully applied to a wide variety of interconnection topologies. These include linear daisy chains (Cheng 88, Bataineh 91, Mani 94), trees (Cheng 90, Bharadwaj 94, Bharadwaj 95, Kim 96), buses (Bataineh 91), hypercubes (Blazewicz 95, Piriyakumar 98) and two and three dimensional meshes (Blazewicz 96, Blazewicz 99, Drozdowski 99, Glazek 03). Figure 5.1, for instance, shows a possible diamond-shaped pattern/flow of load distribution originating from a single processor in a two-dimensional mesh network.

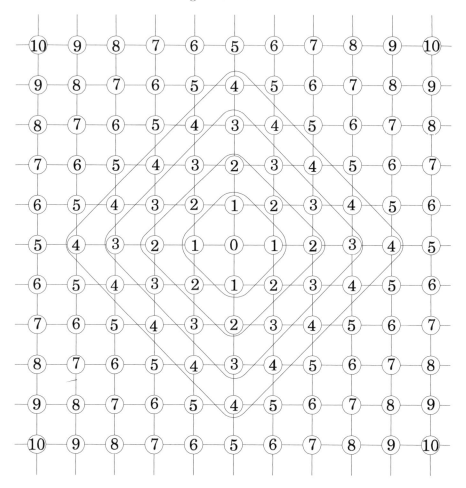

Fig. 5.1. A possible diamond-shaped load distribution flow in a two-dimensional-mesh network. From Blazewicz and Drozdowski © 1996 Foundations of Computing and Decision Sciences

Asymptotic results for infinite-sized networks have also been developed (Ghose 94, Bataineh 97). These are useful as, for sequential load distribution at least, speedup saturates as more processors are added. Therefore if one can guarantee performance close to that of an infinite-sized network with a small to moderate number of processors, one has useful design information.

It should be mentioned that in finding an "optimal" schedule one is doing this in the context of a specific interconnection network and its parameters and in the context of a specific scheduling strategy. In fact, in this chapter it is assumed that the order of load distribution is fixed. In a single level tree with sequential distribution, for instance, this means that there is a predetermined order for children to receive load from the root. Optimization of divisible load schedules for fixed distribution orders is largely an algebraic problem. However one can also optimize the load distribution and solution reporting order using techniques such as simulated annealing, tabu search, genetic programming, or other heuristic algorithms (Charcranoon and Robertazzi 04, Dutot). In some cases analytical results for optimal ordering are possible (Adler 03).

(3) Equivalent Networks: As in other linear theories, such as Markovian queueing theory and resistive electric circuit theory, a complex network can be exactly represented by an exactly "equivalent" network element. For some network topologies such as trees, aggregation can be done recursivley, one subtree of a larger tree at a time.

The manner in which an equivalent processor can be found is straightforward. One finds, using the usual divisible load scheduling techniques, an expression for a subnetwork in a more complex network. For example, such a simple building block may be a pair of adjacent processors and their connecting link in a linear daisy chain network or a single level subtree in a multi-level tree network. One sets the computing speed of a single equivalent processor equal to this subnetwork expression. One then continues this process of aggregating subnetworks of processors, including intermediate equivalent processors, until a single processor is left with a computing speed equivalent to the original network. Final expressions for equivalent processor computing speed may be either closed form or iterative in nature.

(4) Installments and Sequencing: A number of applied optimization problems arise in considering divisible load scheduling. For instance, instead of a node in a tree sequentially distributing load to its children, improved performance results if load is distributed in installments (some to child 1, child 2... child M, more to child 1...) (Bharadwaj 95). Performance under sequential multi-installment load distribution strategies does tend to saturate as the number of installments is increased.

Some sequencing results are surprising. For instance, consider a linear daisy chain network where all processors and links have the same speed. Under one basic sequential scheduling strategy, if load originates at any interior processor, the same solution time results whether load is first distributed to the left or right parts of the network. Other results are more intuitive. For instance, distributing load over a very slow link to a relatively fast processor may degrade overall network solution time (Mani 94).

(5) Scalability: In early studies of divisible load scheduling it was found that if load is distributed from a root node to its children sequentially, as in a tree network, speedup saturates as more nodes are added. Overall the solution time improvement for optimal sequential load distribution over simply dividing the load equally among the processors, if link speed is of the order of processor speed, is on the order of 20% to 40% (Ko 00a). As mentioned, simply increasing the number of installments also suffers from saturated performance as the number of installments is increased. However, recently, it has been found (Hung 02) that if a node transmits load simultaneously to all of its children in a single level tree, speedup is scalable. That is, speedup then grows linearly in the number of children. As long as a node CPU can load output buffers to all links, performance is scalable. Although there was some qualitative sense that this is the case in parallel processing, divisible load theory allows a simple quantitative answer to this problem.

(6) Metacomputing Accounting: A devilish problem in making metacomputing (i.e., distributed computing with payment to computer owners) practical is accounting. That is, how does one take problem size and system parameters into consideration for monetary accounting? Divisible load theory can incorporate an intuitive linear model for computing and communication costs (Sohn 98a). Heuristic rules of simple to moderate complexity can be developed for efficiently (in terms of both cost and performance) assigning load. This can be done in the case where only computation cost is considered or in the case of where both computation and communication costs are considered.

A problem related to meta-computing is parallel processor configuration design. That is, how does one optimally arrange links and processors with certain characteristics (speed and/or cost for instance) in a given topology? Heuristic rules that can be used for the configuration design problem are similar in spirit to those of the metacomputing problem (Charcranoon 00).

(7) Time-Varying Modeling: In actual practice the effort that a computer can devote to a divisible job depends on the status of other background jobs. The same sharing is true of the capacity that a link can provide to transmitting part of the job because of other ongoing transmissions. Figure 5.2 illustrates the time-varying effort received by a single divisible job on a single processor due to "background" jobs sporadically utilizing CPU effort. The upper part of the figure shows background jobs commencing execution (upward arrows) and terminating execution (downward arrows). The lower part of the figure shows the normalized CPU effort available for the single divisible job, which runs for the duration of the diagram. Solution time optimization is possible for divisible loads if the start and end times and effort of such background jobs and messaging is known (Sohn 98b). Integral calculus is used to accomplish this. With less than perfect knowledge of background processes, stochastic modeling can be combined with deterministic divisible load theory.

(8) Unknown System Parameters: Naturally it may be difficult to obtain accurate estimates of available processor effort and link capacity, which are key inputs to divisible load scheduling models. Thus a number of "probing" strategies were recently proposed (Ghose 02) where some small fraction(s) of a load are sent to processors across a network of links for a sort of exploratory processing to allow the estimation of currently available processing capacity at nodes and bandwidth on links.

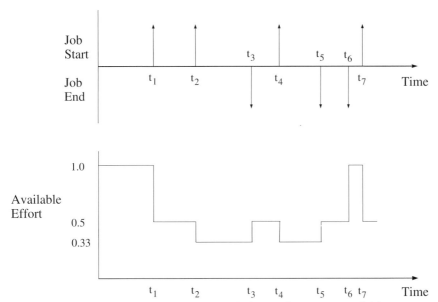

Fig. 5.2. Time-varying background jobs and their influence on available processor effort. From Robertazzi © 2003 IEEE

Although an elegantly simple idea, actual implementations must take several complications into account. These include the time-varying nature of available processor effort and link capacity, the release times of processors (i.e., the times at which processors become free to accept additional load) and assuring that load is distributed on the fastest processors and links. All in all, these probing strategies are a promising approach to robust divisible load scheduling.

(9) Extending Realism: Efforts have been made in recent years to extend the generality of divisible load scheduling. This includes a consideration of systems with finite buffers (Li 00, Drozdowski 03, Bhardawj 03b), finite job granularity (Bharadwaj 00b), start-up costs or fixed charges (Bharadwaj 99, Bharadwaj 00a), scheduling with processor release (i.e., availability) times (Bharadwaj 96a) and scheduling multiple divisible loads (Beaumont 02, Kreaseck 03). Moreover efforts to produce a synthesis of deterministic divisible load modeling and stochastic (queueing) modeling have been made (Sohn 98b, Ko 02, Moges 03). Specialized applications of divisible load scheduling include databases (Ko 00b, Drozdowski 00), image processing, multimedia systems (Bharadwaj 00a) and matrix multiplication (Ghose 98, Kong 01).

(10) Experimental Work: Experiments with actual distributed computer systems show that divisible scheduling theory can be a useful prediction tool (see the DLT experimental work section at the end of this chapter).

IMPLICATIONS

The tractable and realistic nature of divisible load modeling and analysis bodes well for its widespread utility. In a sense this is due to the rich linear mathematics that underly it, as with its cousins, queueing theory and circuit theory. Certainly this flexible analytic structure has been a major reason that these are such large fields. A second important reason, of course, is the breadth of applications. In this context the outlook for future divisible load scheduling theory work and accomplishments is quite promising. The increasing ubiquity of sensor generated data, multiple processor systems, and data intensive computing creates a need for efficient scheduling that should drive further work on theory, applications, and software.

5.2 Some Single Level Tree (Star) Networks

In the following subsections some basic single level tree (star) networks are modeled. For the scheduling policies described, analytical expressions for the optimal allocation of load to each processor as well as the speedup and solution time are found. This modeling and its solutions are from Hung 03b.

We'll distinguish in the following between different types of distribution and the relative start of computation and communication. Under *sequential distribution* load is distributed from a root node to one child at a time. Under *simultaneous distribution* load is distributed from a root to all of its children concurrently. With *staggered start* a child node must receive all of its load before beginning to process load. With *simultaneous start* a node begins processing as soon as it begins to receive load. That is, under simultaneous start a node can receive load and process it at the same time. There are thus four scheduling scenarios with these two sets of possible features. Note that simultaneous distribution was first proposed in divisible load modeling by Piriyakumar and Murthy (98) and simultaneous start was first proposed in divisible load modeling by Kim (03).

A related type of terminology is to note that some processors have front-end sub-processors so that they may compute and communicate at the same time. If a processor does not have a front-end, it can compute or communicate, but not do both at once. So one can, for example, consider a single level tree network with staggered start where the root computes as it distributes load. In this case the root must have a front-end but the children do not have to.

A final piece of terminology sometimes used is to say a root is "intelligent" if it can distribute load as it computes (essentially it has and uses a front-end sub-processor).

The variables we will use in the following are

α_i: The load fraction assigned to the ith link-processor pair.
w_i: The inverse of the computing speed of the ith processor.

z_i: The inverse of the link speed of the ith link.

T_{cp}:   Computing intensity constant:
        The entire load is processed in $w_i T_{cp}$ seconds by the ith processor.

T_{cm}:   Communication intensity constant:
        The entire load can be transmitted in $z_i T_{cm}$ seconds over the ith link.

T_i Finish time of the ith processor.

T_f: The finish time. Time at which the last processor ceases computation.

Note that finish time is called "makespan" in the scheduling literature.

Then $\alpha_i w_i T_{cp}$ is the time to process the fraction α_i of the entire load on the ith processor. Note that the units of $\alpha_i w_i T_{cp}$ are [load] × [seconds/load] × [dimensionless quantity] = [seconds]. Likewise, $\alpha_i z_i T_{cm}$ is the time to transmit the fraction α_i of the entire load over the ith link. Note that the units of $\alpha_i z_i T_{cm}$ are [load] × [seconds/load] × [dimensionless quantity] = [seconds]. The inclusion of T_{cp} and T_{cm} allows the relative communication and computation intensity of a job to be adjusted.

5.2.1 Sequential Load Distribution

Consider a single level tree network where load is distributed sequentially from the root to the children processors, as in Figure 5.3. The root first transmits all of child p_1's load to it, then the root transmits child p_2's load to it, and so on. A child processor starts processing as soon as it begins to receive load. Thus it is assumed that transmission speed is fast enough relative to computation speed that no child processor "starves" for load. Thus we have a case of sequential distribution and simultaneous start. This is a completely deterministic model.

Note that if all of the link speeds are the same, then one has modeled a bus interconnection network.

The process of load distribution can be represented by Gantt chart-like timing diagrams, as illustrated in Figure 5.4. In this Gantt chart-like figure,

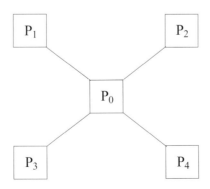

Fig. 5.3. A star (single level tree) interconnection architecture

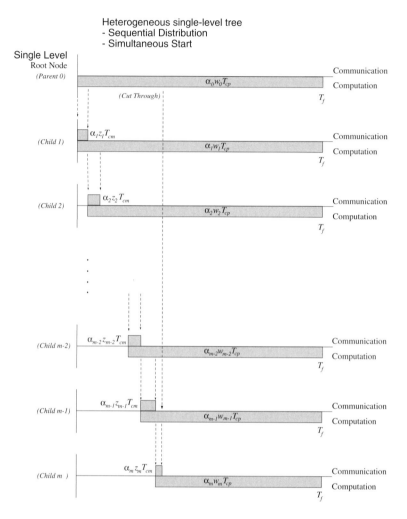

Fig. 5.4. Timing diagram of single level tree with sequential distribution and simultaneous start. From Hung © 2004 IEEE

there is a graph for each processor. The horizontal axis indicates time. Communication is shown above the time axis, and computation is shown below the time axis.

The condition for an optimal solution is that all processors stop processing at the same time. Otherwise load could be transferred from busy to idle processors to improve the solution time (Sohn 96). One can thus write the timing equations as

$$\alpha_0 w_0 T_{cp} = \alpha_1 w_1 T_{cp} \tag{5.1}$$

$$\alpha_1 w_1 T_{cp} = \alpha_1 z_1 T_{cm} + \alpha_2 w_2 T_{cp} \qquad (5.2)$$

$$\alpha_2 w_2 T_{cp} = \alpha_2 z_2 T_{cm} + \alpha_3 w_3 T_{cp} \qquad (5.3)$$

$$.$$
$$.$$
$$.$$

$$\alpha_{m-1} w_{m-1} T_{cp} = \alpha_{m-1} z_{m-1} T_{cm} + \alpha_m w_m T_{cp} \qquad (5.4)$$

Referring to this set of equations and the figure, the first equation equates the processing time of the root to the processing time of child p_1. The second equation equates the processing time of child p_1 to the communication time from the root to child p_1 plus the computation time of child p_2, and so on.

The fundamental recursive equations of the system can be formulated more compactly as follows:

$$\alpha_0 w_0 T_{cp} = \alpha_1 w_1 T_{cp} \qquad (5.5)$$

$$\alpha_{i-1} w_{i-1} T_{cp} = \alpha_{i-1} z_{i-1} T_{cm} + \alpha_i w_i T_{cp} \qquad i = 2, 3, \ldots, m \qquad (5.6)$$

The normalization equation for the single level tree with intelligent root is

$$\alpha_0 + \alpha_1 + \alpha_2 + \cdots + \alpha_m = 1 \qquad (5.7)$$

This gives $m + 1$ linear equations with $m + 1$ unknowns. From equation (5.5)

$$\alpha_0 = \frac{w_1}{w_0} \alpha_1 = \frac{1}{k_1} \alpha_1 \qquad \text{Here, } k_1 = w_0/w_1 \qquad (5.8)$$

From equation (5.6)

$$\alpha_i = \frac{w_{i-1} T_{cp} - z_{i-1} T_{cm}}{w_i T_{cp}} \alpha_{i-1} \qquad (5.9)$$

$$= q_i \alpha_{i-1} \qquad (5.10)$$

$$= \left(\prod_{l=2}^{i} q_l \right) \times \alpha_1 \qquad i = 2, 3, \ldots, m \qquad (5.11)$$

Here, $q_i = (w_{i-1} T_{cp} - z_{i-1} T_{cm})/w_i T_{cp}$. Also, $w_{i-1} T_{cp} > z_{i-1} T_{cm}$. That is, communication time must be faster than computation time; otherwise it is not economical to distribute load. Other conditions for best choosing w_i and z_i appear in Bharadwaj 96b. Then, the normalization equation leads to

$$\alpha_0 + \alpha_1 + \alpha_2 + \cdots + \alpha_m = 1 \qquad (5.12)$$

$$\left[\frac{1}{k_1} + 1 + \sum_{i=2}^{m} \left(\prod_{l=2}^{i} q_l \right) \right] \alpha_1 = 1 \qquad (5.13)$$

$$\alpha_1 = \frac{1}{\frac{1}{k_1} + 1 + \sum_{i=2}^{m}(\prod_{l=2}^{i} q_l)} \tag{5.14}$$

The finish time is

$$T_{f,m} = \alpha_0 w_0 T_{cp} = \frac{1}{k_1}\alpha_1 w_0 T_{cp} \tag{5.15}$$

$$T_{f,m} = \frac{1}{1 + k_1\left[1 + \sum_{i=2}^{m}(\prod_{l=2}^{i} q_l)\right]} w_0 T_{cp}$$

The term $T_{f,m}$ indicates the finish time for the single divisible load solved in a single level tree, which consists of one root node as well as of m children nodes.

The single level tree can be collapsed into a single node, and the inverse of the equivalent computation speed w_{eq} is

$$w_{eq} T_{cp} = T_{f,m} = \frac{1}{1 + k_1\left[1 + \sum_{i=2}^{m}(\prod_{l=2}^{i} q_l)\right]} w_0 T_{cp} \tag{5.16}$$

Then,

$$\gamma_{eq} = \frac{w_{eq}}{w_0} = \frac{1}{1 + k_1\left[1 + \sum_{i=2}^{m}(\prod_{l=2}^{i} q_l)\right]} \tag{5.17}$$

Speedup is the ratio of computation time on one processor to computation time on the entire tree with m children. It is a measure of parallel processing advantage. Since the computation time on a single root processor is

$$T_{f,0} = \alpha_0 w_0 T_{cp} = 1 \cdot w_0 T_{cp} \qquad \alpha_0 = 1, \tag{5.18}$$

the speedup is

$$\boxed{\text{Speedup} = \frac{T_{f,0}}{T_{f,m}} = \frac{1}{\gamma_{eq}} = 1 + k_1\left[1 + \sum_{i=2}^{m}(\prod_{l=2}^{i} q_l)\right]} \tag{5.19}$$

As a special case, consider the situation of a homogeneous network where all children processors have the same inverse computing speed and all links have the same inverse transmission speed (i.e., $w_i = w$ and $z_i = z$ for $i = 1, 2, \ldots, m$). Note that the root w_0 can be different from w_i. Then

$$q_i = \frac{w_{i-1} T_{cp} - z_{i-1} T_{cm}}{w_i T_{cp}} = \frac{w T_{cp} - z T_{cm}}{w T_{cp}} = 1 - \sigma \qquad i = 2, 3, \ldots, m \tag{5.20}$$

Here, $\sigma = z T_{cm}/w T_{cp}$.

Consequently

$$\text{Speedup} = 1 + k_1 \left[1 + \sum_{i=2}^{m} \left(\prod_{l=2}^{i} q_l \right) \right] = 1 + \frac{w_0}{w} \left[\frac{1 - (1 - \sigma)^m}{\sigma} \right] \quad (5.21)$$

If one plots the speedup (or solution time) of this sequential load distribution policy versus the number of children processors, one would see a speedup saturation (approach towards a constant) as the number of children increases. This makes intuitive sense as no matter how many children there are, the root can only distribute load to one child at a time. Thus adding more processors does not significantly improve performance. The calculation of the saturation level for an infinite-sized network appears in a later section. More scalable scheduling strategies appear in the succeeding subsections.

5.2.2 Simultaneous Distribution, Staggered Start

The structure of a single level tree network with intelligent root, $m+1$ processors, and m links is illustrated in Figure 5.5.

If one thinks about it, the performance of the sequential scheduling in the previous section is limited as the root sends load to only one processor at a time. What if the root could send load to all of its children simultaneously? That possibility is discussed in this subsection. Simultaneous transmission of load to all children from the root is possible as long as the CPU is fast enough to continually load buffers for each of its output links.

All children processors are connected to the root processor via direct communication links. The intelligent root processor, assumed to be the only processor at which the divisible load arrives, partitions a total processing load into $m+1$ fractions, keeps its own fraction α_0, and distributes the other fractions $\alpha_1, \alpha_2, \ldots, \alpha_m$ to the children processors respectively and concurrently.

After receiving all of its assigned fraction of load, each processor begins computing immediately (i.e., staggered start) and continues without any interruption until all of its assigned load fraction has been processed. Again, it is assumed that $z_i T_{cm} < w_i T_{cp}$ or communication speed is faster than

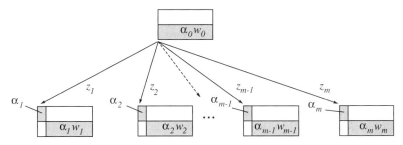

Fig. 5.5. Structure of a single level tree with simultaneous distribution and staggered start

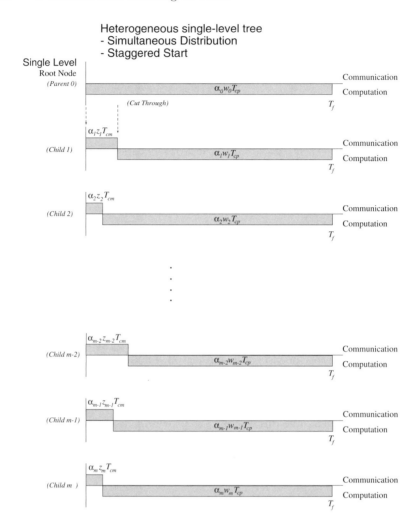

Fig. 5.6. Timing diagram of a single level tree with simultaneous distribution and staggered start

processing speed. In order to minimize the processing finish time, all of the utilized processors in the network must finish computing at the same time (Sohn 96 or Bharadwaj 96b). The process of load distribution can be represented by Gantt chart-like timing diagrams, as illustrated in Figure 5.6. The nodes of Figure 5.5 also contain miniature timing diagrams.

Note, again, that this is a completely deterministic model.

Since for a minimum time solution all processors must stop processing at the same time instant, one can write the fundamental timing equations, using Figure 5.6, as

$$\alpha_0 w_0 T_{cp} = \alpha_1 z_1 T_{cm} + \alpha_1 w_1 T_{cp} \tag{5.22}$$

$$\alpha_1 z_1 T_{cm} + \alpha_1 w_1 T_{cp} = \alpha_2 z_2 T_{cm} + \alpha_2 w_2 T_{cp} \tag{5.23}$$

$$\alpha_2 z_2 T_{cm} + \alpha_2 w_2 T_{cp} = \alpha_3 z_3 T_{cm} + \alpha_3 w_3 T_{cp} \tag{5.24}$$

$$\cdot$$
$$\cdot$$

$$\alpha_{m-1} z_{m-1} T_{cm} + \alpha_{m-1} w_{m-1} T_{cp} = \alpha_m z_m T_{cm} + \alpha_m w_m T_{cp} \tag{5.25}$$

For example the first equation equates the root processing time to the communication time from the root to child processor p_1 plus p_1's processing time. The second equation equates the communication time from the root to p_1 plus the computing time of p_1 to the communication time from the root to p_2 plus the computing time of p_2. The pattern can be naturally generalized.

The normalization equation for the single level tree with intelligent root is

$$\alpha_0 + \alpha_1 + \alpha_2 + \cdots + \alpha_m = 1 \tag{5.26}$$

This yields $m + 1$ linear equations with $m + 1$ unknowns.

Now, one can manipulate the recursive equations to yield the solution.

$$\alpha_0 = \frac{z_1 T_{cm} + w_1 T_{cp}}{w_0 T_{cp}} \alpha_1 = \frac{1}{k_1} \alpha_1 \tag{5.27}$$

Here k_1 is defined as $w_0 T_{cp} / (w_1 T_{cp} + z_1 T_{cm})$. Also

$$\alpha_i = \frac{w_{i-1} T_{cp} + z_{i-1} T_{cm}}{w_i T_{cp} + z_i T_{cm}} \alpha_{i-1} = q_i \alpha_{i-1} \qquad i = 2, 3, \ldots, m \tag{5.28}$$

Here, we designate

$$q_i = \frac{w_{i-1} T_{cp} + z_{i-1} T_{cm}}{w_i T_{cp} + z_i T_{cm}} \tag{5.29}$$

Then, equation (5.28) can be represented as

$$\alpha_i = q_i \alpha_{i-1} = \left(\prod_{l=2}^{i} q_l \right) \alpha_1 \qquad i = 2, 3, \ldots, m \tag{5.30}$$

Employing equations (5.22) and (5.25), the normalization equation becomes

$$\frac{1}{k_1} \alpha_1 + \alpha_1 + \sum_{i=2}^{m} \alpha_i = 1$$

$$\left[\frac{1}{k_1} + 1 + \sum_{i=2}^{m} \left(\prod_{l=2}^{i} q_l \right) \right] \alpha_1 = 1$$

$$\alpha_1 = \frac{1}{\left[\frac{1}{k_1} + 1 + \sum_{i=2}^{m}\left(\prod_{l=2}^{i} q_l\right)\right]} \tag{5.31}$$

From Figure 5.6, the finish time is achieved as

$$T_{f,m} = \alpha_0 w_0 T_{cp} = \frac{1}{k_1}\alpha_1 w_0 T_{cp} \tag{5.32}$$

The term $T_{f,m}$ indicates the finish time for the single divisible load solved in a single level tree, which consists of one root node as well as of m children nodes.

Also, $T_{f,0}$ is defined as the finish time for the entire divisible load processed on the root processor. In other words, $T_{f,0}$ is the finish time of a network composed of only one root node without any children nodes. Hence

$$T_{f,0} = \alpha_0 w_0 T_{cp} = 1 \times w_0 T_{cp} = w_0 T_{cp} \tag{5.33}$$

Now, collapsing a single level tree into a single equivalent node, one can obtain the finish time of the single level tree and the inverse of equivalent computing speed of the equivalent node as follows:

$$T_{f,m} = w_{eq} T_{cp} = \alpha_0 w_0 T_{cp} = \frac{1}{k_1}\alpha_1 w_0 T_{cp} \tag{5.34}$$

Since $\gamma_{eq} = w_{eq}/w_0 = T_{f,m}/T_{f,0}$, one obtains the value of γ_{eq} by equation (5.33) dividing equation (5.34). That is

$$\gamma_{eq} = \frac{1}{k_1}\alpha_1 = \frac{1}{k_1} \times \frac{1}{\left[\frac{1}{k_1} + 1 + \sum_{i=2}^{m}\left(\prod_{l=2}^{i} q_l\right)\right]}$$

$$= \frac{1}{1 + k_1\left[1 + \sum_{i=2}^{m}\left(\prod_{l=2}^{i} q_l\right)\right]} \tag{5.35}$$

Since speedup is the ratio of job solution time on one processor to job solution time on the $m + 1$ processors, one obtains the value of speedup from $T_{f,0}/T_{f,m}$, which is equal to $1/\gamma_{eq}$. Thus,

$$\boxed{\text{Speedup} = \frac{1}{\gamma_{eq}} = k_1 \times \frac{1}{\alpha_1} = 1 + k_1\left[1 + \sum_{i=2}^{m}\left(\prod_{l=2}^{i} q_l\right)\right] \tag{5.36}}$$

Speedup is a measure of the achievable parallel processing advantage. Two cases are discussed for the single level tree below:

1. **General Case:** Since $\prod_{l=2}^{i} q_l$ can be simplified as $(w_1 T_{cp} + z_1 T_{cm})/(w_i T_{cp} + z_i T_{cm})$, the speedup and γ_{eq} can be derived as

$$\text{Speedup} = 1 + k_1 \left[1 + \sum_{i=2}^{m} \left(\prod_{l=2}^{i} q_l \right) \right] \tag{5.37}$$

$$\text{Speedup} = 1 + \frac{w_0 T_{cp}}{w_1 T_{cp} + z_1 T_{cm}} \left[1 + \sum_{i=2}^{m} \frac{w_1 T_{cp} + z_1 T_{cm}}{w_i T_{cp} + z_i T_{cm}} \right]$$

$$\boxed{\text{Speedup} = 1 + w_0 T_{cp} \sum_{i=1}^{m} 1/(w_i T_{cp} + z_i T_{cm})} \tag{5.38}$$

$$\boxed{\gamma_{eq} = 1 / \left(1 + w_0 T_{cp} \sum_{i=1}^{m} \frac{1}{w_i T_{cp} + z_i T_{cm}} \right)} \tag{5.39}$$

2. **Special Case:** As a special case, consider the situation of a homogeneous network where all children processors have the same inverse computing speed and all links have the same inverse transmission speed. In other words, $w_i = w$ and $z_i = z$ for $i = 1, 2, \ldots, m$. Note that the root inverse computing speed, w_0 can be different from those $w_i, i = 1, 2, \ldots, m$. Consequently

$$k_1 = \frac{w_0 T_{cp}}{w_1 T_{cp} + z_1 T_{cm}} = \frac{w_0 T_{cp}}{w T_{cp} + z T_{cm}}$$

$$q_i = \frac{w_{i-1} T_{cp} + z_{i-1} T_{cm}}{w_i T_{cp} + z_i T_{cm}} \qquad i = 2, 3, \ldots, m$$

$$= \frac{w T_{cp} + z T_{cm}}{w T_{cp} + z T_{cm}} = 1$$

$$\gamma_{eq} = \frac{1}{1 + k_1 \left[1 + \sum_{i=2}^{m} \left(\prod_{l=2}^{i} q_l \right) \right]}$$

$$= \frac{1}{1 + \frac{w_0 T_{cp}}{w T_{cp} + z T_{cm}} \left[1 + \sum_{i=2}^{m} \left(\prod_{l=2}^{i} 1 \right) \right]}$$

$$= \frac{1}{1 + \frac{w_0 T_{cp}}{w T_{cp} + z T_{cm}} \left[1 + (m - 1) \right]}$$

$$= \frac{1}{1 + m \times \frac{w_0 T_{cp}}{w T_{cp} + z T_{cm}}} \tag{5.40}$$

$$\boxed{\text{Speedup} = \frac{1}{\gamma_{eq}} = 1 + m \times \frac{w_0 T_{cp}}{w T_{cp} + z T_{cm}} = 1 + k_1 \times m} \qquad (5.41)$$

Our finding is that the computational complexity of the speedup of the single level homogeneous tree, staggered start in this case, is equal to $\Theta(m)$, which is linear in the number of children nodes. Speedup is linear as long as the root CPU can concurrently (simultaneously) transmit load to all of its children. That is, the speedup of the single level tree does not saturate (in contrast to the sequential load distribution of the previous section). Thus the scheduling policy is scalable for this type of network.

5.2.3 Simultaneous Distribution, Simultaneous Start

It would seem reasonable that performance would improve if a child processor could begin processing as soon as the load starts to arrive. Let's consider this scheduling policy.

A single level tree is illustrated in Figure 5.7. All children processors are connected to the root processor via direct communication links. The intelligent root processor partitions a total processing load into $m + 1$ fractions, keeps its own fraction α_0, and distributes the other fractions $\alpha_1, \alpha_2, \ldots, \alpha_m$ to the children processors respectively and concurrently. The children have front ends so load can be received while processing occurs. We assume here that $z_i T_{cm} \ll w_i T_{cp}$, so that the speed of communication in a link is faster than the speed of computation of the processor, which is connected to the link. Therefore, communication ends before computation.

Each processor begins processing the received data while it receives the initial data. In order to minimize the processing finish time, all of the utilized processors in the network should finish computing at the same time.

The process of load distribution can be represented by Gantt chart-like timing diagrams, as illustrated in Figure 5.8.

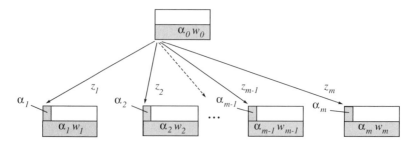

Fig. 5.7. Structure of a single level tree with simultaneous distribution and simultaneous start. From Hung © 2004 ACTA Press

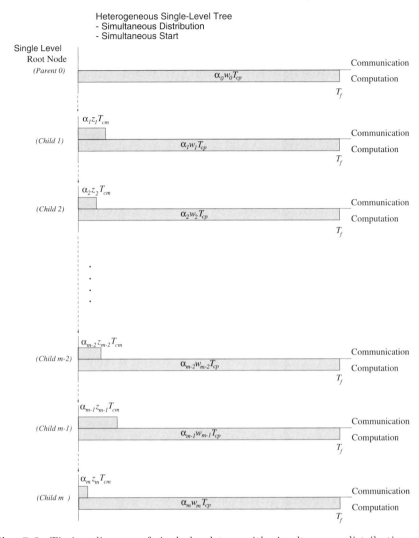

Fig. 5.8. Timing diagram of single level tree with simultaneous distribution and simultaneous start. From Hung © 2004 ACTA Press

Referring to Figure 5.8 the timing equations of this scheduling policy for a single level tree network can be written as

$$\alpha_0 w_0 T_{cp} = \alpha_1 w_1 T_{cp} \tag{5.42}$$

$$\alpha_1 w_1 T_{cp} = \alpha_2 w_2 T_{cp} \tag{5.43}$$

$$\alpha_2 w_2 T_{cp} = \alpha_3 w_3 T_{cp} \tag{5.44}$$

$$.$$

$$.$$

$$\alpha_{m-1} w_{m-1} T_{cp} = \alpha_m w_m T_{cp} \quad i = 2, 3, \ldots, m \tag{5.45}$$

Here the first equation represents equating the processing time of the root processor to the processing time of child processor p_1. The second equation represents equating the processing time of p_1 to the processing time of p_2. These equivalencies can be naturally generalized.

The normalization equation for the single level tree with intelligent root is

$$\alpha_0 + \alpha_1 + \alpha_2 + \cdots + \alpha_m = 1 \tag{5.46}$$

This gives $m + 1$ linear equations with $m + 1$ unknowns.

Now, one can manipulate the recursive equations to yield the solution. From equation (5.42), one obtains

$$\alpha_0 = \frac{w_1 T_{cp}}{w_0 T_{cp}} \alpha_1 = \frac{1}{k_1} \alpha_1 \tag{5.47}$$

Here k_1 is defined as w_0/w_1. From equation (5.42) to equation (5.45), one obtains

$$\alpha_i = \frac{w_{i-1} T_{cp}}{w_i T_{cp}} \alpha_{i-1} = q_i \alpha_{i-1} \quad i = 2, 3, \ldots, m \tag{5.48}$$

Here

$$q_i = \frac{w_{i-1}}{w_i} \tag{5.49}$$

Then

$$\alpha_i = q_i \alpha_{i-1} = \left(\prod_{l=2}^{i} q_l \right) \alpha_1 \quad i = 2, 3, \ldots, m \tag{5.50}$$

Employing equations (5.47) and (5.50), the value of α_1 can be solved by deriving the normalization equation (5.46).

$$\frac{1}{k_1} \alpha_1 + \alpha_1 + \sum_{i=2}^{m} \alpha_i = 1$$

$$\left[\frac{1}{k_1} + 1 + \sum_{i=2}^{m} \left(\prod_{l=2}^{i} q_l \right) \right] \alpha_1 = 1$$

$$\alpha_1 = \frac{1}{\left[\frac{1}{k_1} + 1 + \sum_{i=2}^{m} \left(\prod_{l=2}^{i} q_l \right) \right]} \tag{5.51}$$

Thus, the values of α_0 and α's can be solved with respect to equations (5.47) and (5.50).

From Figure 5.8, the finish time is

$$T_{f,m} = \alpha_0 w_0 T_{cp} = \frac{1}{k_1} \alpha_1 w_0 T_{cp} \tag{5.52}$$

The term $T_{f,m}$ is the finish time of a divisible job solved on the entire tree, consisting of one root node as well as of m children nodes in a single level tree. It is the same as the finish time of an equivalent node for the subtree.

$$T_{f,m} = w_{eq} T_{cp} = \alpha_0 w_0 T_{cp} = \frac{1}{k_1} \alpha_1 w_0 T_{cp} \tag{5.53}$$

Also, $T_{f,0}$ is defined as the solution time for the entire divisible load solved on the root processor. In other words, it is the finish time of a single level tree, composed of only one root node without any child nodes.

$$T_{f,0} = \alpha_0 w_0 T_{cp} = 1 \times w_0 T_{cp} = w_0 T_{cp} \tag{5.54}$$

As before, γ_{eq} is equal to w_{eq}/w_0 or to $T_{f,m}/T_{f,0}$. Thus, from equations (5.53) and (5.54)

$$\gamma_{eq} = \frac{1}{k_1} \alpha_1 = \frac{1}{k_1} \times \frac{1}{\left[\frac{1}{k_1} + 1 + \sum_{i=2}^{m} \left(\prod_{l=2}^{i} q_l \right) \right]}$$

$$= \frac{1}{1 + k_1 \left[1 + \sum_{i=2}^{m} \left(\prod_{l=2}^{i} q_l \right) \right]} \tag{5.55}$$

Based on the definition of speedup

$$\boxed{\text{Speedup} = \frac{1}{\gamma_{eq}} = k_1 \times \frac{1}{\alpha_1} = 1 + k_1 \left[1 + \sum_{i=2}^{m} \left(\prod_{l=2}^{i} q_l \right) \right]} \tag{5.56}$$

Two cases are discussed.

1. **General Case:** From equation (5.49), $\prod_{l=2}^{i} q_l$ can be simplified as w_1/w_i. Therefore, the speedup and γ_{eq} are derived as

$$\text{Speedup} = 1 + k_1 \left[1 + \sum_{i=2}^{m} \left(\prod_{l=2}^{i} q_l \right) \right] \tag{5.57}$$

$$= 1 + \frac{w_0}{w_1} \left[1 + \sum_{i=2}^{m} \frac{w_1}{w_i} \right]$$

$$\text{Speedup} = 1 + w_0 \sum_{i=1}^{m} (1/w_i) \tag{5.58}$$

$$\gamma_{eq} = 1/(1 + w_0 \sum_{i=1}^{m} (1/w_i)) \tag{5.59}$$

2. **Special Case:** As a special case, consider the situation of a homogeneous network where all children processors have the same inverse computing speed and all links have the same inverse transmission speed. In other words, $w_i = w$ and $z_i = z$ for $i = 1, 2, \ldots, m$. Note that the root inverse computing speed w_0 can be different from $w_i, i = 1, 2, \ldots, m$. This will result in

$$k_1 = \frac{w_0}{w_1} = \frac{w_0}{w} \tag{5.60}$$

$$q_i = \frac{w_{i-1}}{w_i} \qquad i = 2, 3, \ldots, m$$

$$= \frac{w}{w} = 1 \tag{5.61}$$

$$\gamma_{eq} = \frac{1}{1 + k_1 \left[1 + \sum_{i=2}^{m} \left(\prod_{l=2}^{i} q_l\right)\right]}$$

$$= \frac{1}{1 + \frac{w_0}{w} \left[1 + \sum_{2}^{m} \left(\prod_{l=2}^{i} 1\right)\right]}$$

$$= \frac{1}{1 + \frac{w_0}{w} \left[1 + (m-1)\right]}$$

$$= \frac{1}{1 + m \times \frac{w_0}{w}} \tag{5.62}$$

$$\text{Speedup} = \frac{1}{\gamma_{eq}} = 1 + m \times \frac{w_0}{w} = 1 + k_1 \times m \tag{5.63}$$

Here, *speedup* is the effective processing gain in using $m+1$ processors. Our finding is that the computational complexity of the speedup of the single level homogeneous tree is equal to $\Theta(m)$, which is proportional to the number of children, per node m. Speedup is a linear function as long as the root CPU can concurrently (simultaneously) transmit load to all of its children. That is, the speedup of the single level tree does not saturate (in contrast to the sequential

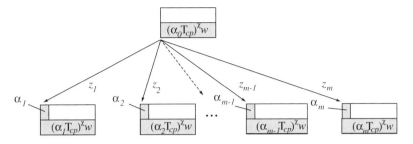

Fig. 5.9. Structure of a single level tree with simultaneous distribution, simultaneous start, and nonlinear computing time

load distribution as in Bharadwaj 96b). It can be seen that this speedup is larger than that of the previous section (simultaneous distribution, staggered start).

Note that one can obtain this speedup result from equation (5.41), the case with staggered start, if one lets $z \to 0$, In other word, the time that the root node distributes load to its children processors is negligible.

5.2.4 Nonlinear Load Processing Complexity

In the three previous subsections, processing time was simply proportional to the load size. This is a linear modeling assumption. But many algorithms have a computing time that is proportional to a nonlinear function of input size. In this section we find the optimal load distribution formula for a power χ dependency between computation time at a node and load size at the node in a single level tree network. All load is available at the root at $t = 0$. The structure of a single level tree network with intelligent root, $m + 1$ processors, and m links is illustrated in Figure 5.9.

All children processors are connected to the root processor via direct communication links. The intelligent root processor, assumed to be the only processor at which the divisible load arrives, partitions a total processing load into $m + 1$ fractions, keeps its own fraction α_0, and distributes the other fractions $\alpha_1, \alpha_2, \ldots, \alpha_m$ to the children processors respectively and concurrently.

While concurrently receiving its initial assigned fraction of load, each processor begins computing immediately and continues without any interruption until all of its assigned load fraction has been processed. This scheduling policy is analogous to the simultaneous distribution, simultaneous start policy of the previous subsection. In order to minimize the processing finish time, all of the utilized processors in the network must finish computing at the same time. The process of load distribution can be represented by a Gantt chart-like timing diagram, as illustrated in Figure 5.10.

Note that this is a completely deterministic model.

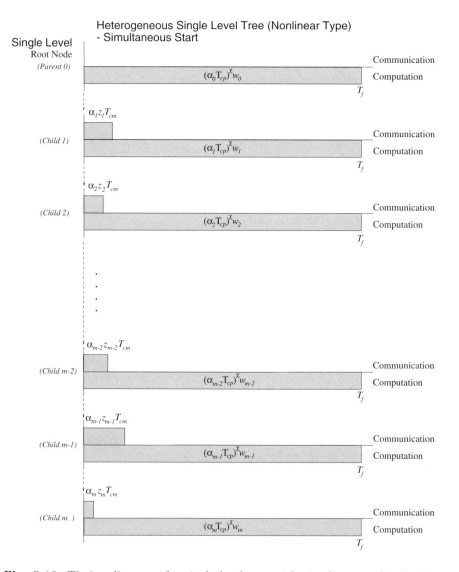

Fig. 5.10. Timing diagram of a single level tree with simultaneous distribution, simultaneous start, and nonlinear computing time

From the timing diagram in Figure 5.10, the fundamental recursive equations of the system can be formulated as follows:

$$(\alpha_0 T_{cp})^{\chi} w_0 = (\alpha_1 T_{cp})^{\chi} w_1 \qquad (5.64)$$

$$(\alpha_1 T_{cp})^{\chi} w_1 = (\alpha_2 T_{cp})^{\chi} w_2 \qquad (5.65)$$

$$(\alpha_2 T_{cp})^\chi w_2 = (\alpha_3 T_{cp})^\chi w_3 \tag{5.66}$$

.

.

$$(\alpha_{m-1} T_{cp})^\chi w_{m-1} = (\alpha_m T_{cp})^\chi w_m \quad i = 2, \ldots, m \tag{5.67}$$

Note that computation time is still linearly proportional to inverse computing speed. The first equation equates the processing time of the root node to the processing time of child processor p_1. The second equation equates the processing time of p_1 to the processing time of p_2, and so on.

The normalization equation for the single level tree with intelligent root is

$$\alpha_0 + \alpha_1 + \alpha_2 + \cdots + \alpha_m = 1 \tag{5.68}$$

This yields $m+1$ linear equations with $m+1$ unknowns. One can manipulate the recursive equations to yield a solution for the optimal allocation of load. From equation (5.64)

$$\alpha_0^\chi = \frac{w_1 T_{cp}^\chi}{w_0 T_{cp}^\chi} \alpha_1^\chi = \frac{w_1}{w_0} \alpha_1^\chi \tag{5.69}$$

Let $\kappa_1 = w_0/w_1$, then equation (5.69) becomes

$$\alpha_0^\chi = \frac{w_1}{w_0} \alpha_1^\chi = \frac{1}{\kappa_1} \alpha_1^\chi \tag{5.70}$$

One obtains (Burrington)

$$\alpha_0 = \sqrt[\chi]{\frac{1}{\kappa_1}} \alpha_1^{\mathrm{hi}} \left(\cos \frac{2k\pi}{\chi} + \sqrt{-1} \, \sin \frac{2k\pi}{\chi} \right) \tag{5.71}$$

where k takes successively the values $0, 1, 2, \ldots, n-1$.

Since $\alpha_0 \geq 0$ and α_0 is *real*, therefore, we take $k = 0$ and get

$$\alpha_0 = \sqrt[\chi]{\frac{1}{\kappa_1}} \cdot \alpha_1 \quad i = 1, 2, \ldots, m \tag{5.72}$$

From equation (5.64) to equation (5.67)

$$\alpha_i^\chi = \frac{w_{i-1} T_{cp}^\chi}{w_i T_{cp}^\chi} \alpha_{i-1}^\chi = \frac{w_{i-1}}{w_i} \alpha_{i-1}^\chi \quad i = 2, \ldots, m \tag{5.73}$$

Let $\xi_i = w_{i-1}/w_i, i = 2, \ldots, m$. Then, equation (5.73) becomes

$$\alpha_i^\chi = \xi_i \alpha_{i-1}^\chi \tag{5.74}$$

One obtains (Burrington)

$$\alpha_i = \sqrt[x]{\xi_i \alpha_{i-1}^{\chi}} \left(\cos \frac{2k\pi}{\chi} + \sqrt{-1} \sin \frac{2k\pi}{\chi} \right) \tag{5.75}$$

where k takes successively the values $0, 1, 2, \ldots, n-1$.

Since $\alpha_i \geq 0$ and α_i is *real*, therefore, we take $k = 0$ and get

$$\alpha_i = \sqrt[x]{\xi_i} \alpha_{i-1} = \sqrt[x]{\prod_{l=2}^{i} \xi_l} \cdot \alpha_1 = \sqrt[x]{\frac{w_1}{w_i}} \cdot \alpha_1 \qquad i = 2, \ldots, m \tag{5.76}$$

Based on the normalization equation (5.68)

$$\alpha_0 + \alpha_1 + \sum_{i=2}^{m} \alpha_i = 1 \tag{5.77}$$

$$\left(\sqrt[x]{\frac{1}{\kappa_1}} + 1 + \sum_{i=2}^{m} \sqrt[x]{\prod_{l=2}^{i} \xi_l} \right) \alpha_1 = 1 \tag{5.78}$$

$$\left(\sqrt[x]{\frac{w_1}{w_0}} + 1 + \sum_{i=2}^{m} \sqrt[x]{\frac{w_1}{w_i}} \right) \alpha_1 = 1 \tag{5.79}$$

$$\left(\sum_{l=0}^{m} \sqrt[x]{\frac{w_1}{w_l}} \right) \alpha_1 = 1 \tag{5.80}$$

$$\sqrt[x]{w_1} \cdot \left(\sum_{l=0}^{m} \sqrt[x]{\frac{1}{w_l}} \right) \alpha_1 = 1 \tag{5.81}$$

Consequently, we obtain

$$\alpha_1 = \frac{1}{\sqrt[x]{w_1} \cdot \left(\sum_{l=0}^{m} \sqrt[x]{\frac{1}{w_l}} \right)} \tag{5.82}$$

From equation (5.72)

$$\alpha_0 = \sqrt[x]{\frac{1}{\kappa_1}} \cdot \alpha_1 = \sqrt[x]{\frac{1}{\kappa_1}} \cdot \frac{1}{\sqrt[x]{w_1} \cdot \left(\sum_{l=0}^{m} \sqrt[x]{\frac{1}{w_l}} \right)}$$

$$= \sqrt[x]{\frac{w_1}{w_0}} \cdot \frac{1}{\sqrt[x]{w_1} \cdot \left(\sum_{l=0}^{m} \sqrt[x]{\frac{1}{w_l}} \right)}$$

$$= \frac{1}{\sqrt[x]{w_0} \cdot \left(\sum_{l=0}^{m} \sqrt[x]{\frac{1}{w_l}} \right)} \tag{5.83}$$

Based on equations (5.76) and (5.82), one obtains

$$\alpha_i = \sqrt[\chi]{\frac{w_1}{w_i}} \cdot \alpha_1 = \sqrt[\chi]{\frac{w_1}{w_i}} \cdot \frac{1}{\sqrt[\chi]{w_1} \cdot \left(\sum_{l=0}^{m} \sqrt[\chi]{\frac{1}{w_l}}\right)} \tag{5.84}$$

$$= \frac{1}{\sqrt[\chi]{w_i} \cdot \left(\sum_{l=0}^{m} \sqrt[\chi]{\frac{1}{w_l}}\right)} \qquad i = 2, \ldots, m \tag{5.85}$$

From equations (5.82), (5.83), and (5.85), we obtain the optimal fractions of load α_i as

$$\alpha_i = \frac{1}{\sqrt[\chi]{w_i} \cdot \left(\sum_{l=0}^{m} \sqrt[\chi]{\frac{1}{w_l}}\right)} \qquad i = 0, 1, 2, \ldots, m \tag{5.86}$$

Using Figure 5.10, the finish time is achieved as

$$T_{f,m} = (\alpha_0 T_{cp})^{\chi} w_0 \tag{5.87}$$

Here, $T_{f,m}$ indicates the finish time for the single divisible load solved in a single level tree, which consists of one root node as well as of m children nodes. Also, $T_{f,0}$ is defined as the finish time for the entire divisible load processed on the root processor. In other words, $T_{f,0}$ is the finish time of a network composed of only one root node without any children nodes. Hence

$$T_{f,0} = (\alpha_0 T_{cp})^{\chi} w_0 = (1 \times T_{cp})^{\chi} w_0 = T_{cp}^{\chi} w_0 \tag{5.88}$$

Now, collapsing a single level tree into a single equivalent node, one can obtain the finish time of the single level tree and the inverse of the equivalent computing speed of the equivalent node as follows:

$$T_{f,m} = (1 \times T_{cp})^{\chi} w_{eq} = T_{cp}^{\chi} w_{eq} = (\alpha_0 T_{cp})^{\chi} w_0 \tag{5.89}$$

As before, $\gamma_{eq} = w_{eq}/w_0 = T_{f,m}/T_{f,0}$ and one obtains the value of γ_{eq} by equation (5.88) dividing equation (5.89). That is

$$\gamma_{eq} = \alpha_0^{\chi} \tag{5.90}$$

Since speedup is the ratio of job solution time on one processor to job solution time on the $m+1$ processors, one obtains the value of speedup from $T_{f,0}/T_{f,m}$, which is equal to $1/\gamma_{eq}$. Thus

$$\text{Speedup} = \frac{1}{\gamma_{eq}} = \left(\frac{1}{\alpha_0}\right)^{\chi} \tag{5.91}$$

$$\boxed{\text{Speedup} = w_0 \left(\sum_{l=0}^{m} \sqrt[\chi]{\frac{1}{w_l}}\right)^{\chi}} \tag{5.92}$$

Speedup is a measure of the achievable parallel processing advantage.

Special Case: As a special case, consider the situation of a homogeneous network where all children processors have the same inverse computing speed and all links have the same inverse transmission speed. In other words, $w_i = w$ and $z_i = z$ for $i = 1, 2, \ldots, m$. Note that the root inverse computing speed w_0 can be different from those $w_i, i = 1, 2, \ldots, m$. From equation (5.83)

$$\alpha_0 = \frac{1}{\sqrt[X]{w_0} \cdot \left(\sum_{l=0}^{m} \sqrt[X]{\frac{1}{w_l}} \right)} \tag{5.93}$$

$$= \frac{1}{\sqrt[X]{w_0} \cdot \left(\sqrt[X]{\frac{1}{w_0}} + \sum_{l=1}^{m} \sqrt[X]{\frac{1}{w}} \right)} \tag{5.94}$$

$$= \frac{1}{\sqrt[X]{w_0} \cdot \left(\sqrt[X]{\frac{1}{w_0}} + m \sqrt[X]{\frac{1}{w}} \right)} \tag{5.95}$$

$$= \frac{1}{1 + m \sqrt[X]{\frac{w_0}{w}}} \tag{5.96}$$

Since $\gamma_{eq} = w_{eq}/w_0 = T_{f,m}/T_{f,0}$, one obtains the value of γ_{eq}

$$\gamma_{eq} = \alpha_0^X = \left(\frac{1}{1 + m \sqrt[X]{\frac{w_0}{w}}} \right)^X \tag{5.97}$$

Since speedup is the ratio of job solution time on one processor to job solution time on the $m + 1$ processors, one obtains the value of speedup from $T_{f,0}/T_{f,m}$, which is equal to $1/\gamma_{eq}$. Thus

$$\boxed{\text{Speedup} = \frac{1}{\gamma_{eq}} = \left(\frac{1}{\alpha_0} \right)^X = \left(1 + m \sqrt[X]{\frac{w_0}{w}} \right)^X} \tag{5.98}$$

Again, speedup is a measure of the achievable parallel processing advantage.

If the computing capability of the root node is the same as that of children nodes for a homogeneous single level tree, i.e., $w_0 = w$, the speedup formula will become

$$\text{Speedup} = (1 + m)^X \tag{5.99}$$

Again, this last result is intuitive.

From the above two equations it can be seen that speedup increases nonlinearly as more and more children processors are added. Are we getting something for nothing in this nonlinear increase? Not really, because as the computational complexity is nonlinear, processing fragments of the load on a number of processors is much more efficient than processing all of the load on one processor.

In fact the situation is indeed a bit more complex. For practical applications involving divisible load with nonlinear computational complexity, in some sense there is some dependency between the data and its processing. That is, once load is divided and processed on separate processors, the individual results from each processor need to be combined through post-processing. This post-processing adds an additional computational cost. The degree to which such divisible processing for loads with nonlinear computational complexity results in an overall computational cost savings is problem dependent.

5.3 Equivalent Processors

The basic model of divisible load scheduling, section 5.2.4 notwithstanding, is a linear one. An important concept in many linear theories, such as linear electric circuits and linear queueing theory, is that of an equivalent element. An equivalent element can replace a subnetwork within a network or replace an entire network and provide exactly identical operating characteristics as the original subnetwork or network, respectively.

Similarly, a network of links and processors operating on a divisible load can be replaced by a single processor with identical processing capability. This is very useful for modeling and analysis. We can calculate the speedup of a large network, for instance, by replacing the network with an equivalent processor and finding the speedup of this equivalent processor.

In the next two subsections, we make this concrete by finding speedup for two types of multi-level tree networks under sequential distribution. The second type has front-end processors at each node. Front-end processors do communication duties for a main processor and thus allow communication to take place at the same time as computation. The first case we look at though is one *without* front-end processors so that communication and computation must take place at different times (the main processor in a node can only do one thing at a time).

Consider a multi-level tree network of communicating processors. In the tree we have three types of nodes (processors): root, intermediate, and terminal nodes. Each tree has one root node that originates the load. An intermediate node can be viewed as a parent of lower level nodes with which it has a direct connection. Also it is a child of an upper level node with which it has a direct connection. The terminal nodes can only be children nodes. The kind and the number of levels in a particular tree determine its size, that is, the total number of nodes in that tree. The type of a symmetrical tree is determined by the number of children nodes that a parent node has. A parent in a "binary" tree would have two children. The root is assumed to be level 0 and its children would be in level 1 and so on. The lowest level is $N - 1$. Every processor can only communicate with its children processors and parent processor. Two adjacent layers of nodes in a tree are referred to as a "level."

In (Cheng 90a) a finite tree, where processors have different (i.e., heterogeneous) speeds, for the above two cases was discussed. However closed form solutions for the minimum finish time were not presented.

In the following we will use the same definitions for α_i, T_{cp}, T_{cm}, w, z, T_i, and T_f as in the previous sections. This material is from Bataineh 94. In this section, the processors in the tree are assumed to all have the same computational speed, $1/w$. The communication speed on a link between a parent processor and each of its children is also assumed to have the same value, $1/z$. These assumptions enable us to tractably collapse the resulting homogeneous tree into one equivalent node that preserves the same characteristics as the original tree. This allows an easy examination of large tree networks. In addition, it becomes possible to find a closed form solution for the optimal amount of data that is to be assigned to each processor in order to achieve the minimum finish time and also to find a numerical solution to the minimum finish time as well as the maximum speedup.

5.3.1 The Tree Network Without Front-End Processors

To collapse the whole tree into one equivalent node we start from the terminal nodes (the last level in the tree, level $N-1$) and move up to the root processor (the first level in the tree, level 0). On our way up, every parent processor and its children will be replaced by one equivalent processor. The process will continue until the root processor and its children are replaced by one equivalent processor. In this aggregation process, only two cases are possible: The first case occurs at the last level where all of the processors have the same speed as shown in Figure 5.11; the second case occurs for the children at level k and their parents at level $k-1$, $k = 1, 2, \ldots, N-2$, where all processors, except the parent, have the same equivalent inverse speed, w_{eq}, as depicted in Figure 5.12. The parent node has inverse processing speed w. In the following, we will discuss analytically the two cases.

The timing diagram of the first case appears in Figure 5.13. Load is distributed to the children processors sequentially. Here each of n homogeneous processors in the network does not contain a front-end processor for communicating off-loading. That is, each processor may either communicate or compute but not do both at the same time. The root processor that originates the load broadcasts to each processor in the network its share of the load before its starts to compute its own share. Each processor begins to compute its share of the load at the moment that it finishes receiving its data. That is, one has here a scheduling policy with sequential distribution, staggered start, and no front-end processors. Note that bus propagation delay is neglected but could be modeled (Blazewicz 97).

Between $t = 0$ and $\alpha_2 Z T_{cm}$ in Figure 5.13, none of the processors performs computation, the first processor communicates data to the second processor, and processors $3, 4, 5, \ldots, n$ are all idle. In general, in the period between $t = 0$ and $t = (\alpha_2 + \alpha_3 + \ldots + \alpha_i) Z T_{cm}$, only $(i-2)$ processors perform computation,

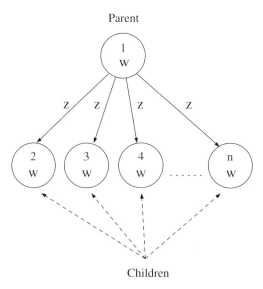

Fig. 5.11. A single level subtree where the children nodes are all terminal nodes in the original multi-level tree

$(n - i)$ processors are idle, $i = 2, 3, \ldots, n$, and two are communicating. This fact serves to increase the minimum finish time.

The equations that relate the various variables and parameters together are stated below

$$T_1 = (1 - \alpha_1)zT_{cm} + \alpha_1 wT_{cp} \qquad (5.100)$$

$$T_2 = \alpha_2 zT_{cm} + \alpha_2 wT_{cp} \qquad (5.101)$$

$$T_3 = (\alpha_2 + \alpha_3)zT_{cm} + \alpha_3 wT_{cp} \qquad (5.102)$$

$$T_4 = (\alpha_2 + \alpha_3 + \alpha_4)zT_{cm} + \alpha_4 wT_{cp} \qquad (5.103)$$

$$T_n = (1 - \alpha_1)zT_{cm} + \alpha_n wT_{cp} \qquad (5.104)$$

The fractions of the total measurement load should sum to one

$$\alpha_1 + \alpha_2 + \cdots + \alpha_n = 1 \qquad (5.105)$$

As mentioned, the minimum finish time is achieved when all processors stop at the same time (Sohn 96), that is, when

$$T_1 = T_2 = T_3 = \cdots = T_n$$

Parent

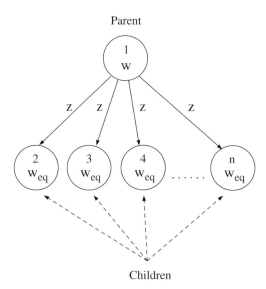

Children

Fig. 5.12. A reduced case where the children are previously collapsed equivalent processors

The originating processor should calculate the optimal values of $\alpha's$. To find these values, one can chain together equations (5.100) through (5.104), and with some algebra, one can write the following equations:

$$\alpha_{n-1} = \alpha_n r \tag{5.106}$$

$$\alpha_3 = \alpha_4 r \tag{5.107}$$

$$\alpha_2 = \alpha_3 r \tag{5.108}$$

$$\alpha_1 = \alpha_n \tag{5.109}$$

Here $r = \frac{wT_{cp} + zT_{cm}}{wT_{cp}}$.

From the above equations the optimal values of $\alpha's$ can be written in terms of α_n and r as follows:

$$\alpha_j = \begin{cases} \alpha_n r^{n-j}, & \text{if } j = 2, 3, \ldots, n-1 \\ \alpha_n, & \text{if } j = 1 \end{cases} \tag{5.110}$$

It is apparent from the above equation that, if the optimal value of α_n can be found, the optimal values of other $\alpha's$ can be readily computed using equation (5.110). Using equations (5.105) and (5.110), one can find the optimal value of α_n in terms of r as follows:

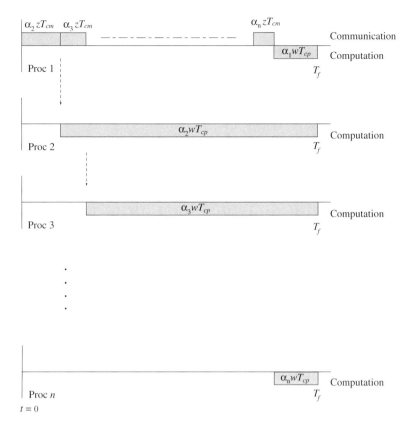

Fig. 5.13. Timing diagram of bottom tree level, no front-end case

$$\alpha_n(1 + r^{n-2} + r^{n-3} + \ldots + r + 1) = 1 \qquad (5.111)$$

$$\alpha_n\left(\sum_{i=1}^{n} r^{n-i} + 1 - r^{n-1}\right) = 1 \qquad (5.112)$$

To solve this we need to solve the summation $\sum_{i=1}^{n} r^{n-i}$. Using a summation formula from the appendix, one has

$$\sum_{i=1}^{n} r^{n-i} = r^n \sum_{i=1}^{n} (1/r)^i$$

$$= r^n\left(\sum_{i=0}^{n} (1/r)^i - 1\right)$$

$$= r^n \left(\frac{1 - (1/r)^{n+1}}{1 - (1/r)} - 1 \right)$$

$$= r^n \left(\frac{1 - (1/r)^{n+1}}{1 - (1/r)} - \frac{1 - (1/r)}{1 - (1/r)} \right)$$

$$= r^n \left(\frac{(1/r) - (1/r)^{n+1}}{1 - (1/r)} \right)$$

$$= \frac{r^{n-1} - (1/r)}{1 - (1/r)}$$

$$= \frac{r^n - 1}{r - 1}$$

Substituting this result into equation (5.112), one has with some algebra

$$\alpha_n = \left(\frac{r - 1}{r^{n-1} + r - 2} \right) \tag{5.113}$$

From equation (5.100) the minimum finish time function T_f for this network architecture is given by

$$T_f = zT_{cm} + \left(\frac{r - 1}{r^{n-1} + r - 2} \right) (wT_{cp} - zT_{cm}) \tag{5.114}$$

Note that for the second term to be positive, communication (zT_{cm}) must be faster than computation (wT_{cp}). Conditions when it is economical to distribute load are discussed in (Bharadwaj 96b). The maximum throughput of the single level tree is

$$\text{Throughput} = \frac{1}{T_f} \tag{5.115}$$

An expression for $w_{eq,t}$ is stated below. Here $w_{eq,t}$ is a constant that is inversely proportional to the speed of an equivalent processor that replaces all the processors in the single level tree of the first case and preserves the same characteristics of the original system. Here the "t" in $w_{eq,t}$ stands for the terminal node case.

$$w_{eq,t} = \frac{1}{T_{cp}} \left(zT_{cm} + \left(\frac{r_t - 1}{r_t^{n-1} + r_t - 2} \right) (wT_{cp} - zT_{cm}) \right) \tag{5.116}$$

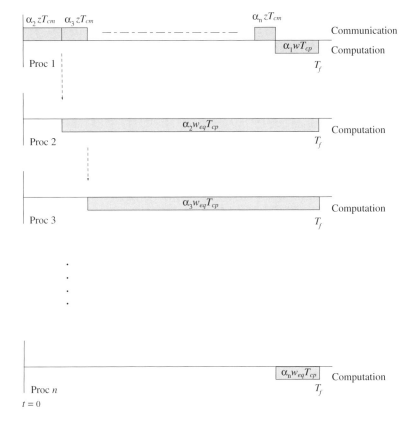

Fig. 5.14. Timing diagram of an intermediate tree level, no front-end case

where

$$r_t = \frac{wT_{cp} + zT_{cm}}{wT_{cp}}$$

This equation is obtained by equating (5.114) and $w_{eq,t}T_{cp}$.

The timing diagram of the second case, a single level tree interior to the multi-level tree, is shown in Figure 5.14 where all processors except the root have the same speed w_{eq}. The time that it takes each processor to process its share is computed by the following set of equations:

$$T_1 = (1 - \alpha_1)zT_{cm} + \alpha_1 wT_{cp} \tag{5.117}$$

$$T_2 = \alpha_2 zT_{cm} + \alpha_2 w_{eq}T_{cp} \tag{5.118}$$

$$T_3 = (\alpha_2 + \alpha_3)zT_{cm} + \alpha_3 w_{eq}T_{cp} \tag{5.119}$$

$$T_4 = (\alpha_2 + \alpha_3 + \alpha_4)zT_{cm} + \alpha_4 w_{eq}T_{cp} \tag{5.120}$$

.
.
.

$$T_n = (1 - \alpha_1)zT_{cm} + \alpha_n w_{eq}T_{cp} \tag{5.121}$$

The fractions of the total measurement load should sum to one

$$\alpha_1 + \alpha_2 + \cdots + \alpha_n = 1 \tag{5.122}$$

The optimal values of $\alpha's$ that has to be assigned to each processor in order to achieve the minimum finish time, based on all processors stopping at the same time, is given by the following set of equations:

$$\alpha_{n-1} = \alpha_n r_i \tag{5.123}$$

$$\alpha_{n-2} = \alpha_{n-1} r_i \tag{5.124}$$

.
.
.

$$\alpha_3 = \alpha_4 r_i \tag{5.125}$$

$$\alpha_2 = \alpha_3 r_i \tag{5.126}$$

$$\alpha_1 = \alpha_n c \tag{5.127}$$

where $r_i = \frac{w_{eq}^i T_{cp} + zT_{cm}}{w_{eq}^i T_{cp}}$ and $c^i = \frac{w_{eq}^i}{w}$

Here the "i" in r_i and w_{eq}^i indicates the ith level subtree. Now the equations can be written in terms of of α_n, r_i, and c^i as follows:

$$\alpha_j = \begin{cases} \alpha_n r_i^{n-j}, & \text{if } j = 2, 3, \ldots, n-1 \\ \alpha_n c^i, & \text{if } j = 1 \end{cases} \tag{5.128}$$

Using equations (5.122) and (5.128), a summation formula from the appendix, and some algebra, α_n can be found as a function of r_i and c. To do this one has

$$\alpha_1 + \alpha_2 + \alpha_3 + \cdots + \alpha_n = = 1$$

$$(\alpha_n c^i + \alpha_n r^{n-2} + \alpha_n r^{n-3} \cdots + \alpha_n r + 1) = 1$$

$$\alpha_n(c^i + r^{n-2} + r^{n-3} + \cdots + r + 1) = 1$$

$$\alpha_n\left(c^i + \sum_{j=1}^{n} r^{n-j} - r^{n-1}\right) = 1$$

Note here that a superscript of c is notational and that of r is a power. But from the solution procedure for equation (5.113), we know that

$$\sum_{j=1}^{n} r^{n-j} = \frac{r^n - 1}{r - 1}$$

Solving algebraically for α_n one has

$$\alpha_n = \frac{r_i - 1}{c^i(r_i - 1) + r_i^{n-1} - 1} \tag{5.129}$$

Now all other optimal values of $\alpha's$ can be computed using equation (5.128). Since $\alpha_1 = \alpha_n c^i$, α_1 can be expressed in terms of r_i and c^i as follows:

$$\alpha_1 = \frac{r_i - 1}{(r_i - 1) + \frac{1}{c^i}(r_i^{n-1} - 1)} \tag{5.130}$$

We now equate equation (5.117) to $w_{eqi}T_{cp}$ in order to find $w_{eq,i}$, a constant that is inversely proportional to the speed of an "equivalent" processor that will replace all processors in Figure 5.12 and preserves the same characteristics as the original system. Note again that for load sharing to produce a net savings, the parenthesis term in equation (5.131) must be positive.

$$w_{eq,i} = z\rho + \alpha_1(w - z\rho) \tag{5.131}$$

where $\rho = T_{cm}/T_{cp}$

Substituting the value obtained for α_1 in the above equation, one finds that

$$w_{eq,i} = z\rho + \frac{r_i - 1}{(r_i - 1) + \frac{1}{c^i}(r_i^{n-1} - 1)}(w - z\rho) \tag{5.132}$$

Starting at level $N - 1$, one can use equation (5.116) to reduce a multi-level tree by one level and then move up one level. Starting from the subtrees whose children are at level $N - 2$ and up to the root processor, one uses equation (5.132) to find $w_{eq_{total}}$. Here $w_{eq_{total}}$ is a constant that is inversely proportional to the speed of an "equivalent" processor that will replace the whole tree while preserving the same characteristics as the original system. Computing $w_{eq_{total}}$, the minimum finish time T_f can be written as follows:

$$T_f = T_{cp}w_{eq_{total}} \tag{5.133}$$

and the multi-level tree throughput is

$$\text{Throughput} = \frac{1}{T_f} \tag{5.134}$$

whereas the maximum speedup is

$$\text{Speedup} = \frac{wT_{cp}}{T_f} = \frac{w}{w_{eq_{\text{total}}}} \tag{5.135}$$

5.3.2 The Tree Network With Front-End Processors

This subsection is similar to the previous one except for the fact that now all the processors in the tree possess front-end processors. That is, each processor can communicate and compute at the same time. This fact will help to reduce the finish time. We will proceed as in the previous subsection and collapse the whole multi-level tree into one equivalent node. We start from the terminal nodes (the last level in the tree, level $N - 1$) and move up to the root processor (the first level in the tree, level 0). As before we will encounter two cases in our aggregation process: The first case occurs at the last two levels where all processors have the same speed; the second case occurs for the children at level k and their parents at level $k - 1$, $k = 1, 2, \ldots, N - 2$, where all processors, except the parent, have the same equivalent speed, w_{eq}, as depicted in Figure 5.11 and Figure 5.12. In the following, we will discuss analytically the two cases.

The timing diagram of the first case is shown in Figure 5.15. Load is, again, distributed sequentially. The root processor that originates the load is now performing both computation and communication simultaneously. Thus, it immediately begins computation on its share of the load while broadcasting the remaining load over the bus to the other processors. Each processor begins to compute its share at the moment that it it finishes receiving its data. Thus we have sequential distribution, staggered start, and processors with front-end processors.

In Figure 5.15 it can be seen that between $t = 0$ and $t = \alpha_2 Z T_{cm}$ the first processor begins to compute its share of the load and communicates with the second processor. All other processors, processors $3, 4, 5 \ldots, n$, are idle. In general, in the period of between $t = 0$ and $t = (\alpha_2 + \alpha_3 + \cdots \alpha_i)Z T_{cm}$; $(n - i)$ processors would be idle; $(i - 1)$ processors perform computation; $i = 2, 3, 4, \ldots, n$; and two are communicating. In the following we will use the same definitions for α_i, w, T_{cp}, z, T_{cm}, T_i, and T_f as in the previous section.

With these definitions, the equations that relate the various variables and parameters together are stated below

$$T_1 = \alpha_1 w T_{cp} \tag{5.136}$$

$$T_2 = \alpha_2 z T_{cm} + \alpha_2 w T_{cp} \tag{5.137}$$

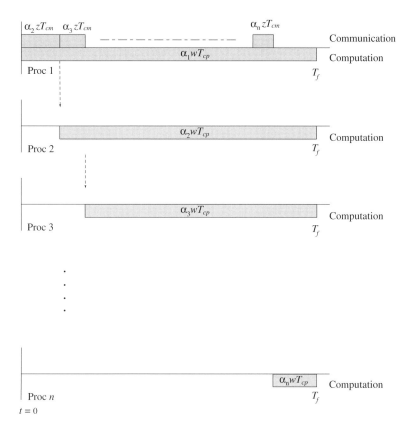

Fig. 5.15. Timing diagram of the bottom tree level, front-end case

$$T_3 = (\alpha_2 + \alpha_3)zT_{cm} + \alpha_3 wT_{cp} \tag{5.138}$$

$$T_4 = (\alpha_2 + \alpha_3 + \alpha_4)zT_{cm} + \alpha_4 wT_{cp} \tag{5.139}$$

$$\vdots$$

$$T_n = (\alpha_2 + \alpha_3 + \cdots + \alpha_n)zT_{cm} + \alpha_n wT_{cp} \tag{5.140}$$

The fractions of the total measurement load should sum to one

$$\alpha_1 + \alpha_2 + \cdots + \alpha_n = 1 \tag{5.141}$$

The objective in analyzing the above equations is to compute the minimum finish time and speedup and compare it with the results that were obtained in the previous sections. The minimum finish time would be achieved when all processors stop at the same time, that is, when

$$T_1 = T_2 = T_3 = \cdots = T_n$$

The optimal values of $\alpha's$ that the originating processor should calculate in order to achieve the minimum finish time can be computed by finding first the following set of equations:

$$\alpha_{n-1} = \alpha_n r \tag{5.142}$$

$$.$$
$$.$$
$$.$$

$$\alpha_3 = \alpha_4 r \tag{5.143}$$

$$\alpha_2 = \alpha_3 r \tag{5.144}$$

$$\alpha_1 = \alpha_2 r \tag{5.145}$$

where

$$r = \frac{wT_{cp} + zT_{cm}}{wT_{cp}} \tag{5.146}$$

Using the above equations the optimal values of the $\alpha's$ can be obtained in terms of α_n and r as shown in the following equation:

$$\alpha_i = \alpha_n r^{n-i} \tag{5.147}$$

where $i = 1, 2, 3, \ldots, n - 1$.

Again, as before, using equations (5.141) and (5.147), one can find α_n in terms of r. The steps to do that are presented in the following equations, which use a summation formula from the appendix (Bataineh 94)

$$\alpha_n(r^{n-1} + r^{n-2} + r^{n-3} + \ldots + r + 1) = 1 \tag{5.148}$$

$$\alpha_n \left(\sum_{i=1}^{n} r^{n-i} \right) = 1 \tag{5.149}$$

$$\alpha_n \left(\frac{r^n - 1}{r - 1} \right) = 1 \tag{5.150}$$

$$\alpha_n = \frac{r - 1}{r^n - 1} \tag{5.151}$$

Knowing the optimal value of α_n, the originating processor can now simply compute all other optimal values of $\alpha's$ by using equation (5.147). The minimum finish time function T_f can be calculated from equation (5.136) (Bataineh 94).

$$T_f = wT_{cp}\frac{r^{n-1}(r-1)}{r^n - 1} \tag{5.152}$$

and the maximum throughput is

$$\text{Throughput} = \frac{1}{T_f} \tag{5.153}$$

These results can be used to obtain an expression for $w_{eq,t}$ which is stated below. Here $w_{eq,t}$ is a constant that is inversely proportional to the speed of an equivalent processor that replaces all the processors in the single level tree of the first case and preserves the characteristics of the original system. Again, the "t" in $w_{eq,t}$ stands for the terminal node case.

$$w_{eq,t} = w\frac{r_t^{n-1}(r_t - 1)}{r_t^n - 1} \tag{5.154}$$

where

$$r_t = \frac{wT_{cp} + zT_{cm}}{wT_{cp}}$$

This equation is obtained by equating equation (5.152) and $w_{eq,t}T_{cp}$.

The timing diagram of the second case, a single level tree interior to the multilevel tree is shown in Figure 5.16. The time that it takes each processor to process its share is computed by the following set of equations:

$$T_1 = \alpha_1 wT_{cp} \tag{5.155}$$

$$T_2 = \alpha_2 zT_{cm} + \alpha_2 w_{eq}T_{cp} \tag{5.156}$$

$$T_3 = (\alpha_2 + \alpha_3)zT_{cm} + \alpha_3 w_{eq}T_{cp} \tag{5.157}$$

$$T_4 = (\alpha_2 + \alpha_3 + \alpha_4)zT_{cm} + \alpha_4 w_{eq}T_{cp} \tag{5.158}$$

$$\cdot$$
$$\cdot$$
$$\cdot$$

$$T_n = (1 - \alpha_1)zT_{cm} + \alpha_n w_{eq}T_{cp} \tag{5.159}$$

The fractions of the total measurement load should sum to one

$$\alpha_1 + \alpha_2 + \cdots + \alpha_n = 1 \tag{5.160}$$

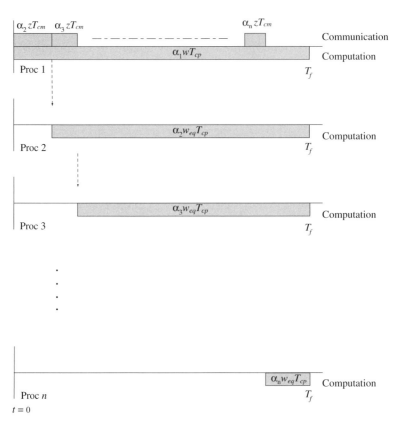

Fig. 5.16. Timing diagram of an intermediate tree level, front-end case

The optimal values of $\alpha's$ that have to be assigned to each processor in order to achieve the minimum finish time is given by the following set of equations:

$$\alpha_{n-1} = \alpha_n r_i \qquad (5.161)$$

$$\alpha_{n-2} = \alpha_{n-1} r_i \qquad (5.162)$$

$$\vdots$$

$$\alpha_3 = \alpha_4 r_i \qquad (5.163)$$

$$\alpha_2 = \alpha_3 r_i \qquad (5.164)$$

$$\alpha_1 = \alpha_2 c \qquad (5.165)$$

where $r_i = \dfrac{w^i_{eq}T_{cp}+zT_{cm}}{w^i_{eq}T_{cp}}$ and $c^i = \dfrac{w^i_{eq}T_{cp}+zT_{cm}}{wT_{cp}}$.

Here i indicates the level of children nodes being considered. It should be noted that, to achieve the minimum finish time, α_i was solved for by equating T_i to T_{i+1}. The equations can be written in terms of of α_n, r_i, and c^i as follows:

$$\alpha_j = \begin{cases} \alpha_n r^{n-j}, & \text{if } j = 2, 3, \ldots, n-1 \\ \alpha_n r_i^{n-2} c^i, & \text{if } j = 1 \end{cases} \qquad (5.166)$$

Using equations (5.160) and (5.166), α_n can be found as a function of r_i and c. The steps are

$$\alpha_1 + \alpha_2 + \alpha_3 + \cdots + \alpha_n = = 1$$

$$(\alpha_{n-2}r^{n-2}c^i + \alpha_n r^{n-2} + \alpha_n r^{n-3} \cdots + \alpha_n r + 1) = 1$$

$$\alpha_n(r^{n-2}c^i + r^{n-2} + r^{n-3} + \cdots + r + 1) = 1$$

$$\alpha_n\left(r^{n-2}c^i + \sum_{j=1}^{n} r^{n-j} - r^{n-1}\right) = 1$$

Note here, again, that the superscript of c is notational and that of r is a power. But from the solution procedure for equation (5.113) we know that

$$\sum_{j=1}^{n} r^{n-j} = \frac{r^n - 1}{r - 1}$$

Using algebra one can find α_n as

$$\alpha_n = \frac{r_i - 1}{(c^i + 1)r_i^{n-1} - c^i r_i^{n-2} - 1} \qquad (5.167)$$

Now all other optimal values of $\alpha's$ can be computed using equation (5.166). Since $\alpha_1 = \alpha_n r_i^{n-2} c$, α_1 can be expressed in terms of r_i and c as follows:

$$\alpha_1 = \frac{r_i - 1}{c^i(r_i^{n-1} - r_i^{n-2}) + r_i^{n-1} - 1}(r_i^{n-2}c^i) \qquad (5.168)$$

$$= \frac{r_i^{n-1} - r_i^{n-2}}{r_i^{n-1} - r_i^{n-2} + \frac{1}{c^i}(r_i^{n-1} - 1)}$$

In order to find $w_{eq,i}$, we equate equation (5.155) to $w_{eq,i}T_{cp}$. Here $w_{eq,i}$ is a constant that is inversely proportional to the speed of an "equivalent" processor that will replace all processors in the single level tree of the second case and preserves the characteristics as the original system. Again, the "i" indicates the level of children nodes being considered.

$$w_{eq,i} = w\alpha_1 \tag{5.169}$$

Substituting the value obtained for α_1 in the above equation, we find that

$$w_{eq,i} = w\left(\frac{r_i^{n-1} - r_i^{n-2}}{r_i^{n-1} - r_i^{n-2} + \frac{1}{c^i}(r_i^{n-1} - 1)}\right) \tag{5.170}$$

Starting at level $N - 1$, one can use equation (5.154) to reduce the multi-level tree by one level and then move up to level $N - 2$. Starting from the subtrees where children are at level $N - 2$ and up to the root processor, one uses equation (5.170) to find $w_{eq_{total}}$. Here $w_{eq_{total}}$ is a constant that is inversely proportional to the speed of an "equivalent" processor that will replace the whole tree while preserving the same characteristics as the original system. Computing $w_{eq_{total}}$, the minimum finish time T_f can be written as follows:

$$T_f = w_{eq_{total}} T_{cp} \tag{5.171}$$

and the throughput is

$$\text{Throughput} = \frac{1}{T_f} \tag{5.172}$$

whereas the maximum speedup is

$$\text{Speedup} = \frac{wT_{cp}}{T_f} = \frac{w}{w_{eq_{total}}} \tag{5.173}$$

5.4 Infinite-Sized Network Performance

5.4.1 Linear Daisy Chains

A linear daisy chain of processors where processor load is divisible and shared among the processors will be examined in this subsection. It is shown, as in the previous section, that two or more processors can be collapsed into a single equivalent processor. This equivalence allows a characterization of the nature of the minimal time solution, a simple method to determine when to distribute load for linear daisy chain networks of processors without front-end communication subprocessors and closed form expressions for the equivalent processing speed of infinitely large daisy chains of processors.

The situation to be considered involves a linear daisy chain of processors, as is illustrated in Figure 5.17. A single "problem" (or job) is solved on the network at one time. It takes time $w_i T_{cp}$ to solve the entire problem on processor i. Here w_i is inversely proportional to the speed of the ith processor and T_{cp} is the normalized computation intensity when $w_i = 1$. It takes time $z_i T_{cm}$

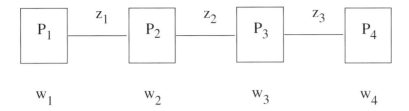

Fig. 5.17. Linear daisy chain network

to transmit the entire problem representation (data) over the ith link. Here z_i is inversely proportional to the channel speed of the ith link and T_{cm} is the normalized communication intensity when $z_i = 1$.

It is assumed that the problem representation can be divided among the processors in a divisible fashion. That is, fraction α_i of the total problem is assigned to the ith processor so that its computing time becomes $\alpha_i w_i T_{cp}$. It is desired to determine the optimal values of the α_i's so that the problem is solved in the minimum amount of time. The situation is nontrivial as communication delays are incurred in transmitting fractional parts of the problem representation to each processor from the originating processor.

Two cases will be considered: processors that have front-end communications subprocessors for communications off-loading so that communication and computation may proceed simultaneously and processors without front-end communications subprocessors so that communication and computation must be performed at separate times.

A timing diagram for a linear daisy chain network of four processors with front-end communications subprocessor (as in Figure 5.17) is illustrated in Figure 5.18. It is assumed that the problem (load) originates at the left-most processor.

At time 0, processor 1 can start working on its fraction α_1 of the problem in time $\alpha_1 w_1 T_{cp}$. It also simultaneously communicates the remaining fraction of the problem to processor 2 in time $(\alpha_2 + \alpha_3 + \alpha_4)z_1 T_{cm}$. Processor 2 can then begin computation on its fraction of the problem (in time $\alpha_2 w_2 T_{cp}$) and communicates the remaining load to processor 3 in time $(\alpha_3 + \alpha_4)z_2 T_{cm}$. The process continues until all processors are working on the problem. Note that the store and forward switching method is used here, but other protocols could be modeled as well.

A similar, but not identical, situation for a linear daisy chain network with processors that do not have front-end communications subprocessors is illustrated in Figure 5.19. Here each processor must communicate the remaining load to its right neighbor before it can begin computation on its own fraction.

In (Cheng 88) recursive expressions for calculating the optimal α_i's were presented. These are based on the simplifying premise that, for an optimal allocation of load, all processors must stop processing at the same time. Intuitively

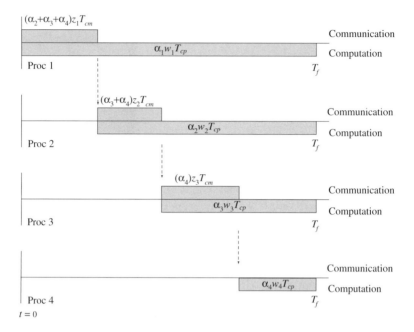

Fig. 5.18. Network with front-end communications subprocessors

this is because otherwise some processors would be idle while others were still busy. Analogous solutions have been developed for tree networks (Cheng 90) and bus networks (Bataineh 91, Bharadwaj 96b).

In this subsection the concept of collapsing two or more processors and associated links into a single processor with equivalent processing speed is presented. This allows a complete proof [an abridged one appears in (Cheng 88)] that for the optimal, minimal time solution all processors must stop at the same time. Moreover, for the case without front-end communications subprocessors, it allows a simple algorithm, to determine when it is economical to distribute load among multiple processors. Finally, the notion of equivalent processors will enable the derivation of simple closed form expressions for the equivalent speed of a linear daisy chain network containing an infinite number of processors. This provides a limiting value for the performance of this network architecture and scheduling policy.

Equivalent Processors

Consider a linear daisy chain network of N processors as in Figure 5.17. Two adjacent processors may be combined into a single "equivalent" processor that presents operating characteristics to the rest of the network that are identical to those of the original two processors. Two cases, processors with and without front-end communications subprocessors, will be considered (Robertazzi 93).

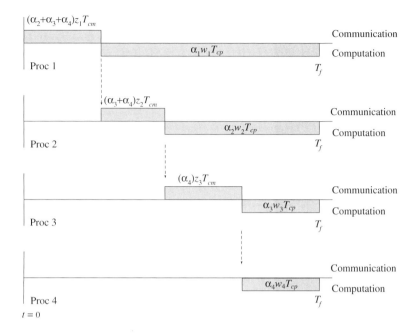

Fig. 5.19. Network without front-end communications subprocessors

In both cases it is assumed that the load originates at the left-most processor (processor 1). If the load originates at an interior processor one can use the same methodology to collapse the processors to the left and the right of the originating processor into equivalent processors and then collapse the remaining three processors into a single equivalent processor (Bataineh 97).

Let's first consider processors with front-end communications subprocessors. We will start with the N-1st and Nth (right most) of N processors, as illustrated in Figure 5.20. The figure begins at the moment when load has finished being transmitted to the N-1st processor from the N-2nd processor. As in (Cheng 88), the N-1st processor keeps $\hat{\alpha}_{N-1}$ fraction of what it receives and transmits the remaining $1 - \hat{\alpha}_{N-1}$ fraction to the Nth processor. It should be mentioned that the use of hatted variables is not the only way to find the results below. The total load received by the N-1st processor from the N-2nd is $(\alpha_{N-1} + \alpha_N)$. The time each is active, from the figure, is

$$T_{N-1} = \hat{\alpha}_{N-1}(\alpha_{N-1} + \alpha_N)w_{N-1}T_{cp} \tag{5.174}$$

$$T_N = (1 - \hat{\alpha}_{N-1})(\alpha_{N-1} + \alpha_N)z_{N-1}T_{cm}$$
$$+ (1 - \hat{\alpha}_{N-1})(\alpha_{N-1} + \alpha_N)w_N T_{cp} \tag{5.175}$$

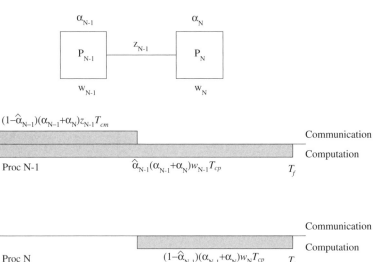

Fig. 5.20. Timing diagram for the N-1st and Nth processors with front-end communications subprocessors

To prove that the minimal time solution occurs when both processors stop at the same time, the possibilities $T_{N-1} \geq T_N$ and $T_{N-1} \leq T_N$ must be examined. If $T_{N-1} \geq T_N$, simple algebra results in

$$\hat{\alpha}_{N-1} \geq \frac{z_{N-1}T_{cm} + w_N T_{cp}}{w_{N-1}T_{cp} + z_{N-1}T_{cm} + w_N T_{cp}} \tag{5.176}$$

Here equality occurs when both processors stop at the same time. Minimizing the solution time $T_{sol} = T_{N-1}$ clearly requires

$$\min T_{sol} = (\min \hat{\alpha}_{N-1})(\alpha_{N-1} + \alpha_N)w_{N-1}T_{cp} \tag{5.177}$$

This is so that the optimal value of $\hat{\alpha}_{N-1}$ occurs for equality in equation (5.176). The quantity $(\alpha_{N-1} + \alpha_N)$ is not involved in the minimization since the value of $\hat{\alpha}_{N-1}$ is unaffected by the total load, $(\alpha_{N-1} + \alpha_N)$, delivered to the N-1st processor. Put another way, the optimization involves the fraction of load being allocated between P_{N-1} and P_N, not the total load allocated to these two processors. The other half of the proof, for $T_{N-1} \leq T_N$, is similar.

The two processors with front-end (fe) communications subprocessors can be replaced by a single processor with equivalent inverse speed constant

$$w_{eq}^{fe} = \frac{\hat{\alpha}_{N-1}(\alpha_{N-1} + \alpha_N)w_{N-1}T_{cp}}{(\alpha_{N-1} + \alpha_N)T_{cp}} = \hat{\alpha}_{N-1}w_{N-1} \tag{5.178}$$

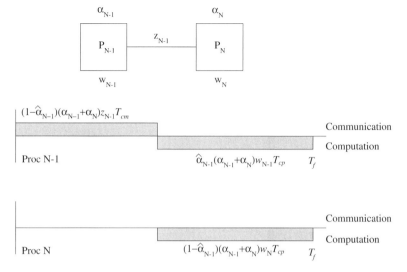

Fig. 5.21. Timing diagram for the N-1st and Nth processors without front-end communications subprocessors

Here $\hat{\alpha}_{N-1}$ is given by equation (5.176) with equality. The solution time is divided by the normalized computation time to yield the equivalent inverse speed constant. Thus, starting with the N-1st and Nth processors, the entire linear chain of processors can be collapsed, two at a time, into a single equivalent processor. Thus one can recursively show that, for a network of N processors, the optimal solution occurs when all processors stop at the same time.

Now let's examine a linear daisy chain where the processors have no front-end subprocessors. Again, consider the N-1st and Nth (right most) of N processors in a linear chain. Figure 5.21 starts from the moment when load has finished being transmitted from the N-2nd to the N-1st processor. Again, as in (Cheng 88), the N-1st processor keeps $\hat{\alpha}_{N-1}$ fraction of what it receives and transmits the remaining $1 - \hat{\alpha}_{N-1}$ fraction to the Nth processor. From Figure 5.21, the time each is active is

$$T_{N-1} = (1 - \hat{\alpha}_{N-1})(\alpha_{N-1} + \alpha_N)z_{N-1}T_{cm}$$
$$+ \hat{\alpha}_{N-1}(\alpha_{N-1} + \alpha_N)w_{N-1}T_{cp} \qquad (5.179)$$

$$T_N = (1 - \hat{\alpha}_{N-1})(\alpha_{N-1} + \alpha_N)z_{N-1}T_{cm}$$
$$+ (1 - \hat{\alpha}_{N-1})(\alpha_{N-1} + \alpha_N)w_N T_{cp} \qquad (5.180)$$

Once again, to prove that the minimal time solution requires both processors to stop at the same time, the cases $T_{N-1} \geq T_N$ and $T_{N-1} \leq T_N$ can be considered. For $T_{N-1} \geq T_N$, simple algebra results in

$$\hat{\alpha}_{N-1} \geq \frac{w_N}{w_{N-1} + w_N} \tag{5.181}$$

Here equality occurs when both processors stop at the same time. From equation (5.179), the solution time can be rewritten as

$$T_{sol} = T_{N-1} = (\alpha_{N-1} + \alpha_N)z_{N-1}T_{cm}$$

$$+ \hat{\alpha}_{N-1}(\alpha_{N-1} + \alpha_N)(w_{N-1}T_{cp} - z_{N-1}T_{cm}) \tag{5.182}$$

The sign of the term $(w_{N-1}T_{cp} - z_{N-1}T_{cm})$ now becomes important. If it is positive, minimizing T_{sol} is equivalent to minimizing $\hat{\alpha}_{N-1}$ and the optimal solution occurs at equality for equation (5.181). In other words, if $w_{N-1}T_{cp} > z_{N-1}T_{cm}$, communication is fast enough relative to computation that the distribution of load is economical. Again, $(\alpha_{N-1} + \alpha_N)$ is not involved in the minimization.

On the other hand, if $(w_{N-1}T_{cp} - z_{N-1}T_{cm})$ is negative, then minimizing T_{sol} is equivalent to maximizing $\hat{\alpha}_{N-1}$ at $\hat{\alpha}_{N-1} = 1$. That is, communication speeds are slow relative to computation speed so that it is more economical for processor N-1 to process the entire load itself rather than to distribute part of it to processor N.

The case where $T_{N-1} \leq T_N$ proceeds along similar lines. Again, the ability to collapse processors into equivalent processors allows one to extend the proof that two processors must stop at the same time for a minimal time solution to N processors.

When to Distribute Load

A practical problem for the case without front-end communications subprocessor is to compute the equivalent computation speed of a linear daisy chain network when, in fact, the optimal solution may not make use of all processors, because of too slow communication speeds. Again, if the load originates at the left-most processor, this can be done by collapsing the processors, two at a time, from right to left in Figure 5.17, into a single equivalent processor. However, when looking at two adjacent processors, say the i-1st and the ith (where the ith is an equivalent processor for processors $i, i+1, \ldots$), one must determine whether it is economical to distribute load. That is, one seeks the faster of either the solution with both processors T_{both} or with just the single i-1st processor T_{single}

$$T_{\text{both}} = (1 - \hat{\alpha}_{i-1})(\alpha_{i-1} + \alpha_i)z_{i-1}T_{cm}$$

$$+ \hat{\alpha}_{i-1}(\alpha_{i-1} + \alpha_i)w_{i-1}T_{cp} \tag{5.183}$$

$$T_{\text{single}} = (\alpha_{i-1} + \alpha_i)w_{i-1}T_{cp} \tag{5.184}$$

Here fraction $\hat{\alpha}_{i-1}$ of the total load $(\alpha_{i-1} + \alpha_i)$ is assigned to processor $i - 1$ and fraction $1 - \hat{\alpha}_i$ is assigned to processor i. If $T_{\text{single}} < T_{\text{both}}$ then the ith processor is removed from consideration and the equivalent processing speed constant, with no front-end (nfe) communications subprocessor, is

$$w_{eq}^{nfe} = \frac{(\alpha_{i-1} + \alpha_i)w_{i-1}T_{cp}}{(\alpha_{i-1} + \alpha_i)T_{cp}} = w_{i-1} \tag{5.185}$$

If $T_{\text{single}} > T_{\text{both}}$ then load distribution is economical and the two processors are collapsed into a single equivalent processor with speed constant

$$w_{eq}^{nfe} = \frac{(1 - \hat{\alpha}_{i-1})z_{i-1}T_{cm} + \hat{\alpha}_{i-1}w_{i-1}T_{cp}}{T_{cp}} \tag{5.186}$$

From equation (5.181)

$$\hat{\alpha}_{i-1} = \frac{w_i}{w_{i-1} + w_i} \tag{5.187}$$

Note that in equation (5.186) factors of $(\alpha_{i-1} + \alpha_i)$ cancel in the numerator and the denominator.

By keeping track of which of equations (5.183) and (5.184) is smaller, it is possible to determine which processors to remove from the final network.

Note that the above procedure can also be applied to the situation when the load originates at a processor, which is located in the interior of the network. The parts of the network to the left and to the right of the originating processor can be collapsed, into equivalent processors, following the previous procedure. The remaining three processors (left, originating, right) can then be further collapsed into a single equivalent processor. Naturally, it must be checked whether the inclusion of the left and/or right equivalent processor leads to a faster solution.

Infinite Number of Processors

A difficulty with the linear network daisy-chained architecture is that as more and more processors are added to the network, the amount of improvement in the network's equivalent speed approaches a saturation limit. Intuitively, this is because of the overhead in communicating the problem representation down the linear daisy chain in what is essentially a store and forward mode of operation.

It is possible to develop simple expressions for the equivalent inverse processing speed of an infinite number of homogeneous processors and links.

These provide a limiting value on the performance of this architecture. The technique is similar to that used for infinitely sized electrical networks to determine equivalent impedance.

Let the load originate at a processor at the left boundary of the network (processor 1). The basic idea is to write an expression for the speed of the single equivalent processor for processors $1, 2 \ldots \infty$. This is a function of the speed of the single equivalent processor for processors $2, 3 \ldots \infty$. However these two speeds should be equal since both involve an infinite number of processors. One can simply solve the resulting implicit equation for this speed.

Consider, first, the case where each processor has a front-end communications subprocessor. Let $w_i = w$ and $z_i = z$. Let the network consist of P_1 and an equivalent processor for processors $2, 3 \ldots \infty$. Then

$$w_{eq}^{fe} = \hat{\alpha}_1 w \tag{5.188}$$

But from equation (5.176) with equality, and making the above assumption

$$w_{eq}^{fe} = \frac{z\rho + w_{eq}^{fe}}{w + z\rho + w_{eq}^{fe}} w \tag{5.189}$$

Here $\rho = T_{cm}/T_{cp}$. Solving for w_{eq}^{fe} results in

$$w_{eq}^{fe} = \left(-z\rho + \sqrt{(z\rho)^2 + 4wz\rho}\right)/2 \tag{5.190}$$

The solution time for such an infinite network is simply given by $T_{sol} = w_{eq}^{fe} T_{cp}$.

In a similar manner, an expression for the equivalent processing speed of a linear daisy chain network with an infinite number of processors with no front-end (nfe) communications subprocessors can be determined. Again, the load originates at processor 1 at the left boundary of the daisy chain.

$$w_{eq}^{nfe} = \sqrt{wz\rho} \tag{5.191}$$

The solution time for this infinite network is simply given by $T_{sol} = w_{eq}^{nfe} T_{cp}$.

This last expression is somewhat intuitive. Doubling w and z doubles w_{eq}^{nfe}. Doubling either w or z alone increases w_{eq}^{nfe} by a factor of $\sqrt{2}$. These results agree very closely with numerical results presented in (Cheng 88). It is straightforward to show that $w_{eq}^{fe} < w_{eq}^{nfe}$. Thus, in this limiting case, solution time is always reduced through the use of front-end processors.

It is also possible to use the above results to calculate the limiting performance of an infinite-sized daisy chain when the load originates at a processor at the interior of the network (with the network having infinite extent to the left and the right). Expressions (5.190) or (5.191) can be used to construct equivalent processors for the parts of the network to the left and right of the originating processor. The resulting three-processor system then can be simply solved.

The concept of collapsing two or more processors into an equivalent processor has been shown to be useful in examining a variety of aspects related to these linear daisy chain networks of load sharing processors. Expressions for the performance of infinite chains of processors are particularly useful as if one can construct a finite-sized daisy chain that approaches the performance of a hypothetical infinite system, one can feel comfortable that performance can not be improved further for this particular architecture and load distribution sequence.

5.4.2 Tree Networks

In this subsection load distribution for networks with a tree topology with sequential load distribution is discussed (Bataineh 97). This material is more general than a simple consideration of hard-wired tree networks of processors. This is because a natural way to distribute load in a processor network with cycles is through the use of an embedded spanning tree.

A homogeneous binary tree network of communicating processors will be considered in this subsection. The general technique developed here can be applied to other types of symmetrical and homogeneous tree networks. In the tree there are three types of processors: root, intermediate, and terminal processors. Each tree has one root processor that originates the load. An intermediate processor can be viewed as a parent of lower level processors with which it has a direct connection. Also it is a child of an upper level processor with which it has a direct connection. The terminal processors can only be children processors.

Every processor can only communicate with its children processors and parent processor. Each of the processors in the tree is assumed to have the same computational speed $1/w$. The communication speed between a parent processor and each of its children is also assumed to have the same value $1/z$.

In this section a binary tree where processors are equipped with front-end processors for communications off-loading will be discussed. Therefore communication and computation can take place in each processor at the same time.

In Cheng 90 a finite tree for the above case was discussed. The minimum processing time is achieved when all processors in the tree stop at the same time. Moreover formal proofs of optimality of single level trees are available (Sohn 96). As the size of the tree becomes larger, the share assigned to the root processor becomes smaller and so the processing time decreases. On the other hand, adding more processor (node) levels to the tree, will result in more overhead time spent in communicating small fractions of load to the

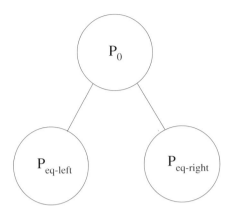

Fig. 5.22. Three-node reduced tree network

new processors. At some point, adding more processors will not decrease the fractions of load assigned to the root processor substantially and so there is not a considerable improvement in the processing time. In that case, it may be advisable not to add more processors (hardware) to the tree since the cost of doing so may not be worth the small improvement in the performance of the system.

The idea behind obtaining the asymptotic processing time for this tree where $n = \infty$ is to collapse the tree into three processors as shown in Figure 5.22. The right side of the tree below the root P_0 has been replaced by one "equivalent" processor with equivalent processing speed w^∞_{eq}. The same is true for the left side of the tree below the root P_0 where it was replaced with one "equivalent" processor that has an equivalent computational speed w^∞_{eq}. Naturally, as the left and right subtrees are homogeneous infinite trees in their own right, an equivalent processor for either one of them has the same computational speed as one for the entire tree.

The timing diagram for this equivalent system that preserves the characteristics of an infinite-sized binary tree is shown in Figure 5.23. From this figure it can be seen that the computing time of the root processor $\alpha_0 w T_{cp}$ equals the communication time between the parent processor (root processor) and the left processor $\alpha_l z T_{cm}$ plus the computing time of the left equivalent processor $\alpha_l w^\infty_{eq} T_{cp}$. Also the computing time of the left-side equivalent processor $\alpha_l w^\infty_{eq} T_{cp}$ equals the communication time between the root processor and the right side "equivalent" processor $\alpha_r z T_{cm}$ plus the computing time of the right equivalent processor $\alpha_r w^\infty_{eq} T_{cp}$. If the three processors in Figure 5.22 are replaced with one equivalent processor, then the computing time of the root processor $\alpha_0 w T_{cp}$ equals the computing time of the equivalent processor $w^\infty_{eq} T_{cp}$. The three equations just explained are listed below

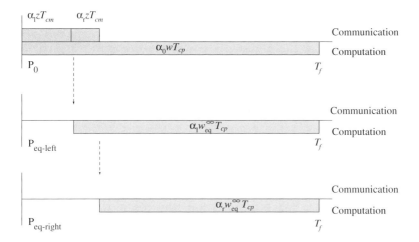

Fig. 5.23. Timing diagram for Figure 5.22 network (with front-end subprocessors case)

$$\alpha_0 w T_{cp} = \alpha_l z T_{cm} + \alpha_l w_{eq}^\infty T_{cp} \tag{5.192}$$

$$\alpha_l w_{eq}^\infty T_{cp} = \alpha_r z T_{cm} + \alpha_r w_{eq}^\infty T_{cp} \tag{5.193}$$

$$\alpha_0 w T_{cp} = w_{eq}^\infty T_{cp} \tag{5.194}$$

Also the sum of the fractions of the load equals one

$$\alpha_0 + \alpha_r + \alpha_l = 1 \tag{5.195}$$

Now, there are four equations with four unknowns, namely w_{eq}^∞, α_0, α_r, and α_l. With some algebra one can show that w_{eq}^∞ can be determined by iteratively solving the equation

$$w_{eq}^\infty = \frac{w(z T_{cm} + w_{eq}^\infty T_{cp})}{z T_{cm} + w_{eq}^\infty T_{cp} + w T_{cp} + \frac{w w_{eq}^\infty T_{cp}^2}{z T_{cm} + w_{eq}^\infty T_{cp}}} \tag{5.196}$$

Solving this equation is equivalent to solving the following cubic equation:

$$(w_{eq}^\infty)^3 + [2z\rho + w](w_{eq}^\infty)^2 - [z\rho(w - z\rho)]w_{eq}^\infty - w z^2 \rho^2 = 0 \tag{5.197}$$

Consequently, the ultimate finish time for an infinite tree network with front-end processors T_{fe}^∞ can now be computed by

$$T_{fe}^\infty = w_{eq}^\infty T_{cp} \tag{5.198}$$

Naturally

$$\text{Speedup} = \frac{w}{w_{eq}^{\infty}} \tag{5.199}$$

In a similar way the solution time for a homogeneous tree without front ends can be found. This technique does not apply to heterogeneous (i.e., networks with different link and processor speeds) infinite networks. It can be extended, though, to a symmetrical network with three or more children per node.

5.5 Time-Varying Environments

All of the previous sections investigated divisible loads under the assumption that a processor can compute only a single job at a time. Under this assumption, the next job can be served only after the processor finishes the computation of the currently running job. However, most practical time-sharing computer systems can handle more than one job at a time. It is therefore natural to study divisible loads in multiprogrammed and multiprocessor environments (Sohn 98b).

In the previous sections processor and link speed were constant. Let's consider situations where they may vary with time. The processors here can be assumed to be multiprogrammed so that there are a number of jobs running in the *background* in addition to a divisible load of interest. These background jobs consume processor and link resources so that the divisible load of interest may see time-varying processor and link speed. It is immaterial for our purposes whether the background jobs are divisible or indivisible. The processor speed and the channel speed depend on the number of jobs that is currently served under a processor or transmitted through a channel. When there are a large number of jobs running in a processor, the processor speed for a specific job of interest becomes slower than when it has fewer jobs. The channel speed also becomes slower when there are a large number of background job related transmissions passing through a link than when there are fewer transmissions using the link.

The purpose of this section is to determine the optimal fraction of the entire workload to be distributed to each processor to achieve the minimal processing time when the processor speed of the processors are a time-varying variables. To determine the optimal fraction of the workload deterministically, the processor speed over the duration of the divisible load computation must be known in advance before the load originating processor starts distributing the workload to each processor. If the exact arrival times and departure times of the background jobs are known, one can determine the exact time-varying processor speed and the channel speed. This is suitable for *production jobs* that are performed in a system repeatedly for a known period. If the arrival and the departure times of the background jobs are not known, but the stochastic

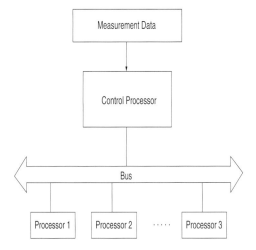

Fig. 5.24. Bus network with load origination at control processor

arrival process and the stochastic departure process of the jobs can be assumed to be Markovian, the optimal fraction of the workload can still be found by a stochastic analysis that makes use of well-known Markovian queueing theory. In this section a deterministic numerical method to find the optimal allocation of the entire workload in terms of minimal processing time is presented when the background jobs' arrival and departure times are known.

Time-Varying Processor Speed

The distributed computing system to be considered here consists of a control processor for distributing the workload and N processors attached to a linear bus as in Figure 5.24. New arriving measurement data are distributed to each processor under the supervision of the control processor. The control processor distributes the workload among the N processors interconnected through a bus type communication medium in order to obtain the benefits of parallel processing. Note that the control processor is a network processor that does no processing itself and only distributes the workload.

Each processor is a multiprogrammed processor that can simultaneously process multiple jobs. Thus the processor speed varies with time and it depends on the amount of workload. The processor speed varies under the following processor sharing rule: The processor devotes all its computational power evenly to each job. That is, if there are m jobs running under a certain processor, each job receives $1/m$ of the full computational power of the processor. This behavior is similar to a fair resource scheduling policy as used in UNIX systems. It is assumed here that there is no limitation of the number of jobs to be simultaneously processed in a single processor, even though the

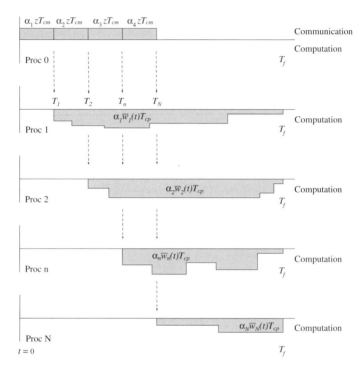

Fig. 5.25. Timing diagram for bus network with time-varying processor speed

processor speed for a specific job will be very slow if there are a large number of jobs running simultaneously under the processor.

In this section we assume that background jobs start and terminate simultaneously across all processors and that negligible bus communication is needed to support their running. The technique can be extended to situations with different background job start/stop times on different processors. The main problem in this paper is to find the optimal fraction of a divisible load, which is distributed to each of N processors to minimize the processing finish time when the communication delay is nonnegligible.

The timing diagram for the bus network with load origination at a control processor is depicted in Figure 5.25. In this timing diagram, communication time appears above the horizontal time axis and computation time appears below the axis. In this section, the channel speed is assumed to be a constant, whereas the computing speed of each processor is assumed to be time varying, as described above.

At any time, the processor effort available for the divisible loads of interest varies because of background jobs that consume processor effort. These background jobs can arrive at or terminate on the processors at any time during the computation of the divisible load that the control processor is going to distribute. The arrival and departure times of the background jobs over intervals during which the divisible load is processed, however, should be exactly known. This is the reason that this section represents deterministic models of the load sharing problem. When the arrival and departure times are unknown and the statistics of the arrival and departure process of the jobs are known to be Markovian, then this load sharing problem can be stochastically analyzed as in (Sohn 98b).

Referring to Figure 5.25, at time $t = 0$, the originating processor (the control processor in this case) transmits the first fraction of the workload to P_1 in time $\alpha_1 z T_{cm}$. The control processor then sequentially transmits the second fraction of the workload to P_2 in time $\alpha_2 z T_{cm}$, and so on. After P_1 completes receiving its workload from the control processor (an amount of α_1 of the entire load), P_1 can start computing immediately and it will take a time of $T_f - T_1$ to finish. Here $T_1 = \alpha_1 z T_{cm}$. The second processor P_2 also completes receiving the workload from the control processor at time $T_2 = (\alpha_1 + \alpha_2) z T_{cm}$, and it will start computing for a duration of $T_f - T_2$ of time. This procedure continues until the last processor. For optimality, all processors must finish computing at the same time. Intuitively, this is because otherwise the processing time could be improved by transferring the load from busy processors to idle ones.

Now let us represent those intervals of the computation time as $T_f - T_1, T_f - T_2, \ldots, T_f - T_N$. The interval $T_f - T_n$ for P_n to compute the nth fraction of the entire load can be expressed as

$$T_f - T_n = \alpha_n \overline{w}_n(t) T_{cp} \qquad\qquad n = 1, 2, \ldots, N \qquad (5.200)$$

where $\overline{w}_n(t)$ is defined as the inverse of the time average of the applied computing speed of P_n in the interval (T_n, T_f). Since $w_n(t)$ is defined as *the inverse* of the computing speed, to calculate the time average of $w_n(t)$ one must invert $w_n(t)$ first to make it proportional to the actual computing speed and take the time average, and then invert it again. That is

$$\overline{w}_n(t) = \left(E\left\{ \frac{1}{w_n(t)} \right\} \right)^{-1}$$

$$= \frac{T_f - T_n}{\int_{T_n}^{T_f} \frac{1}{w_n(t)} dt} \qquad\qquad (5.201)$$

Explanatory diagrams for the computing speed of P_n are depicted in Figure 5.26(a), (b), and (c). Consider Figure 5.26(a), (b), and (c) in reverse order. Figure 5.26(c) shows the process that is *proportional* to the computing speed of P_n, which is available for the single divisible job of interest. The divisible job arrives at time 0. When the processor is idle in the interval (t_0, t_1), the divisible load that is delivered from the control processor will receive the full computational power of P_n. Therefore, the computing speed of P_n in the interval (t_0, t_1) for the load from the control processor is $1/w_n$, where w_n is the *inverse* of the maximum computational power of P_n. When there is one background job running in the processor in the interval (t_1, t_2) due to the arrival of one background job at time $t = t_1$, the computational power of P_n is equally divided by two so that each job, one background job and the divisible load from the control processor, can receive half of the full computational power of P_n. That is, the computing speed of P_n in the interval (t_1, t_2) for each job is $1/2w_n$.

Likewise, when there are two background jobs running in the processor in the interval (t_2, t_3) due to the additional arrival of a background job at time $t = t_2$, the computational power of P_n is equally divided by three so that each job, two background jobs and the divisible load from the control processor, can receive one third of the full computational power of P_n. The computing speed of P_n in the interval (t_2, t_3) for each job is $1/3w_n$. When the processor finishes the computation of one of the background jobs at time $t = t_3$, the computing speed of the P_n for each job (at this time, there are two jobs running in the processor, one a background job and the other a divisible load fragment from the control processor) speeds up back to $1/2w_n$.

Note that the integral in the denominator of equation (5.201) is the area under the curve of Figure 5.26(c) between times T_n and T_f.

Figure 5.26(b) shows the process that is *inversely proportional* to the computing speed of P_n. In other words, Figure 5.26(b) is just the inverse of

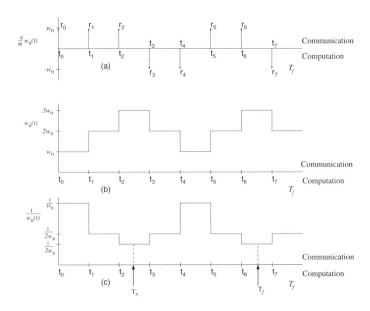

Fig. 5.26. (a) Derivative of timing process that is inversely proportional to computing speed. (b) Timing process that is inversely proportional to computing speed. (c) Timing process that is proportional to computing speed

Figure 5.26(c). Figure 5.26(a) is the derivative of Figure 5.26(b). This represents the arrival and departure time of the jobs. The upright impulse r_0 indicates the arrival of a divisible job that is present in the system for the duration of the timing diagram. The upright impulses (r_1, r_2, r_5, r_6) represent the arrival of each background job, and the upside down impulses (r_3, r_4, r_7) represent the departure or service completion of each background job.

What is deterministic in this section is that the time of each arrival and departure of the background jobs is deterministically known. That is, the time $t_1, t_2, \ldots$, etc. should be all known at time $t = 0$. This condition can be true of a production system repetitively running the same jobs. The height of the each impulse is $+w_n$ for the ones that correspond to the arrivals and $-w_n$ for the ones that correspond to the departure of the background job. This is because one arrival of a background job causes the computing speed to change from $1/mw_n$ to $1/(m+1)w_n$ in Figure 5.26(c) so the inverse speed changes from mw_n to $(m+1)w_n$ in Figure 5.26(b) for any integer m. A similar explanation can applied to the departure of background jobs.

Let us now find the expressions for Figure 5.26(a), (b), and (c). The expression for Figure 5.26(a) is

$$\frac{d}{dt} w_n(t) = \sum_{k=0}^{\infty} r_k \delta(t - t_k) w_n \tag{5.202}$$

where

$$r_k = \begin{cases} +1, \text{ for arrival} \\ -1, \text{ for departure} \end{cases}$$

The following equation represents Figure 5.26(b):

$$w_n(t) = \sum_{k=0}^{\infty} r_k u(t - t_k) w_n \tag{5.203}$$

Here $u(t)$ is the unit step function (i.e., a function that has the value of 1 for positive time and has a value of 0 for negative time). A little thought yields an expression for Figure 5.26(c)

$$\frac{1}{w_n(t)} = \sum_{k=0}^{\infty} \left(\sum_{j=0}^{k} r_j \right)^{-1} [u(t - t_k) - u(t - t_{k+1})] \frac{1}{w_n} \tag{5.204}$$

The next step is to find the time average of $w_n(t)$ in the interval (T_n, T_f). To find $\overline{w}_n(t)$, it is necessary to find $\int_{T_n}^{T_f} \frac{1}{w_n(t)} dt$ from equation (5.201).

$$\int_{T_n}^{T_f} \frac{1}{w_n(t)} dt = \frac{T_f}{w_n(T_f)} - \frac{T_n}{w_n(T_n)} - \sum_{k=x_n+1}^{x_f} \left(\frac{1}{w_n(t_k)} - \frac{1}{w_n(t_{k-1})} \right) t_k \tag{5.205}$$

See (Sohn 98b) for details. Therefore,

$$\overline{w}_n(t) = \frac{T_f - T_n}{\frac{T_f}{w_n(T_f)} - \frac{T_n}{w_n(T_n)} - \sum_{k=x_n+1}^{x_f} \left(\frac{1}{w_n(t_k)} - \frac{1}{w_n(t_{k-1})}\right) t_k} \tag{5.206}$$

Using equation (5.200), one can also find the expression for α_n.

$$T_f - T_n = \alpha_n \overline{w}_n(t) T_{cp}$$

$$= \alpha_n T_{cp} \frac{T_f - T_n}{\int_{T_n}^{T_f} \frac{1}{w_n(t)} dt} \tag{5.207}$$

Thus

$$\alpha_n = \frac{1}{T_{cp}} \int_{T_n}^{T_f} \frac{1}{w_n(t)} dt$$

$$= \frac{1}{T_{cp}} \left[\frac{T_f}{w_n(T_f)} - \frac{T_n}{w_n(T_n)} - \sum_{k=x_n+1}^{x_f} \left(\frac{1}{w_n(t_k)} - \frac{1}{w_n(t_{k-1})}\right) t_k \right]$$

$$\tag{5.208}$$

Here equations (5.205), (5.206), and (5.208) are functions of T_n and T_f. That is, if T_n and T_f are known, the fraction of the workload for P_n as well as the integral of the applied computing speed of the nth processor and the inverse of the average applied computing speed of P_n in the interval (T_n, T_f) can be found. This problem can be solved by a simple recursive method that can express every α_n as a function of T_f. Let us introduce an algorithm to find the optimal fraction of workload that the control processor must calculate before distributing the load to each processor.

1. Express α_N as a function of T_f from

$$\alpha_N = \frac{1}{T_{cp}} \int_{T_N}^{T_f} \frac{1}{w_N(t)} dt$$

   Since $T_N = (\alpha_1 + \alpha_2 + \cdots + \alpha_N)ZT_{cm} = ZT_{cm}$, T_N is known.
2. Express α_{N-1} as a function of T_f from

$$\alpha_{N-1} = \frac{1}{T_{cp}} \int_{T_{N-1}}^{T_f} \frac{1}{w_{N-1}(t)} dt$$

   Since $T_{N-1} = (1 - \alpha_N)ZT_{cm}$, T_{N-1} is a function of α_N and is a function of T_f.
3. Express α_{N-2} as a function of T_f from

$$\alpha_{N-2} = \frac{1}{T_{cp}} \int_{T_{N-2}}^{T_f} \frac{1}{w_{N-2}(t)} dt$$

Since $T_{N-2} = (1 - \alpha_N - \alpha_{N-1})ZT_{cm}$, T_{N-2} is a function of α_N and α_{N-1} and is a function of T_f.

4. This procedure can be continued up to α_1. Then, one can express every α_n as a function of T_f. Finally, by using the normalization equation that states that $\sum_{n=1}^{N} \alpha_n = 1$, all of the α_n, as well as the actual T_f, can be found.

Note that the algorithm starts from time 0 when the initial processor speeds are known as they are a function of past arrivals and departures.

5.6 Linear Programming and Divisible Load Modeling

In this chapter divisible load scheduling models have been solved by linear equation solution and related recursions. An alternative solution technique is to use linear programming. Linear programming (Hillier, Robertazzi 99) is an optimization technique for constrained linear models developed originally by George Dantzig in 1947. Generally one has a number of continuous variables, a linear objective function of the variables that must be minimized or maximized along with a set of linear constraint equations in the variables.

The simplex algorithm of George Dantzig is an effective means of solving such linear models. It can be shown that the feasible solutions of a linear programming problem lie on the extreme (corner) points of a convex polyhedron in the solution space. One might intuitively and loosely think of the convex polyhedron as a faceted diamond. The simple algorithm moves from extreme point to adjacent extreme point until the extreme point corresponding to the optimal solution is reached. A different interior point method was developed by Narendra Karmarkar in 1984. Interior point methods start at a point inside the convex polyhedron and move outward until the extreme point corresponding to the optimal solution is reached. Agrawal and Jagadish in 1988 were the first to show that linear programming could be used to solve divisible load models.

As an example, consider a divisible load distribution model for a single level tree with simultaneous distribution and staggered start, as in section 5.2.2. From Figure 5.6. one can write expressions for the finish time, T, as

$$\alpha_0 w_0 T_{cp} \leq T \tag{5.209}$$

$$\alpha_1 z_1 T_{cm} + \alpha_1 w_1 T_{cp} \leq T \tag{5.210}$$

$$\alpha_2 z_2 T_{cm} + \alpha_2 w_2 T_{cp} \leq T \tag{5.211}$$

$$\cdot$$

$$\alpha_i z_i T_{cm} + \alpha_i w_i T_{cp} \leq T \tag{5.212}$$

$$\alpha_m z_m T_{cm} + \alpha_m w_m T_{cp} \leq T \tag{5.213}$$

These are the constraints (i.e., the time that each processor finishes communication and computation is less than or equal to the system finish time T).

The objective function is

$$\min T \tag{5.214}$$

But

$$T = \alpha_0 w_0 T_{cp} \tag{5.215}$$

So the complete linear program is

$$\min \alpha_0 w_0 T_{cp} \tag{5.216}$$

$$\alpha_1 z_1 T_{cm} + \alpha_1 w_1 T_{cp} - \alpha_0 w_0 T_{cp} \leq 0 \tag{5.217}$$

$$\alpha_2 z_2 T_{cm} + \alpha_2 w_2 T_{cp} - \alpha_0 w_0 T_{cp} \leq 0 \tag{5.218}$$

.

$$\alpha_i z_i T_{cm} + \alpha_i w_i T_{cp} - \alpha_0 w_0 T_{cp} \leq 0 \tag{5.219}$$

.

$$\alpha_m z_m T_{cm} + \alpha_m w_m T_{cp} - \alpha_0 w_0 T_{cp} \leq 0 \tag{5.220}$$

$$\alpha_0 + \alpha_1 + \alpha_2 \cdots + \alpha_m - 1 = 0 \tag{5.221}$$

$$\alpha_0, \alpha_1, \alpha_2 \ldots, \alpha_m \geq 0 \tag{5.222}$$

Linear programming is more computation intensive than an analytical solution. The advantage of using it (or using a linear equation solution to some extent) is that it takes less analytical effort to simply write the linear program (or set of linear equations) and solve it with a library function than to solve the model analytically. This is particularly advantageous for a complex model. On the other hand an analytical solution often gives more intuition into the nature of the solution and is faster to solve, once found. Thus the two approaches are complimentary.

5.7 Experimental Work

The first experiments known to the author on optimally distributing divisible load were reported by Agrawal and Jagadish at Bell Laboratories in 1988. The experiments involved nine AT&T 3B2/300 computers running over an Ethernet compatible network. Problems run involved determining the Ramanujam number (i.e., find the smallest N such that $N = a^3 + b^3$ and $N = c^3 + d^3$ where $a \neq b \neq c \neq d$). A more challenging problem as it requires both appreciable communication and computation that was studied is matrix multiplication (see also Ghose and Kim 98). Both equal and optimal division of load were considered. An "excellent match" between theory and experiment was found.

Work by M. Drozdowski and colleagues at the Poznan University of Technology in Poland since 1994 has investigated the accuracy and predictability of the DLT model (Drozdowski 00). The investigations focused on comparing the real execution time of applications with the predictions of DLT.

Experiments were conducted on transputer systems (simple dedicated platforms where the user has total system control). Experiments on networks of workstations (NOWs) used Suns, IBM SP-2s, and PCs with PVM, MPI, and Java. The test parallel applications were search for a pattern in a text file, compression of a file, join of databases, and graph coloring by the use of a genetic search metaheuristic.

In the transputer system, the difference between the predictions of the model and experimental data was in the range of 1%, due to the simplicity of the computing system. The results for NOWs present a more complex picture. The difference between theory and practice ranged from 1%, to as much as 40%. The best accuracy was observed when computations were long and large volumes of data were transferred in the communications. In such cases the linear part of the computation and communication times dominated.

As computer and communication speeds have increased over the years, the minimum size of a divisible job whose processing can be accurately predicted by the divisible load scheduling theory has commensurately increased. For relatively short amounts of divisible data, other phenomena come into light: nondeterministic execution time of the operating system services, dependence of computation and communication speeds on the size of the load, and the possibly nonuniform nature of the load.

Figure 5.27 illustrates the difference between modeling and measurements for a relational database "join" operation on six 133 MHz PCs in a 1999 experiment (Drozdowski 00). Here the horizontal axis represents job size (in bytes). Here also FIFO indicates results are returned from processors to the originating (root) processor in the same order that load was received. Under LIFO results are returned in the opposite order. Note that the accuracy of the divisible load models increases as job size $V[B]$ increases.

Other work by B. Veeravalli and colleagues at the National University of Singapore since 1998 has focused on implementing the scheduling algorithms proposed in the DLT literature to real-life situations that naturally

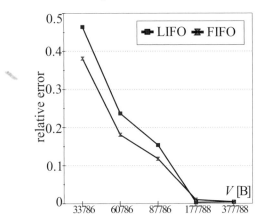

Fig. 5.27. Difference between modeling and measurement for join type relational database operation

qualify under the domain of embarrassingly parallel computations. Problems investigated to date include low-level processing of images for edge-detection applications, large-scale matrix-vector product computations and processing electro-magnetic field strength computations for CAD applications.

For edge-detection applications (Bharadwaj 02), the performance of load distribution strategies recommended by DLT were tested on both the workstation clusters (HP machines) under PVM platforms and PC clusters (comprising high-speed Pentium series machines). Scheduling large-size image processing under resource constraints has been studied. Images of size ranging from 512×512 to 3000×3000, typical of satellite pictures, have been processed.

In the case of large-scale matrix-vector product computations (Kong 01), a matrix of size $200 \times 100,000$, typical of several industrial applications including the design of industrial microwave ovens and conducting finite element methods for large-scale mechanical engineering applications, has been implemented on a PC cluster. Here, a distributed software architecture has been developed that carries out the load distribution on a bus network. Parallelization strategies have also been developed for SGI machines for computing electro-magnetic field strengths around a given circuit layout. This study aids CAD designers to tune the layout as per the interference levels between any pair of copper strips in a design.

Veeravalli and colleagues have found the match between theory and experiment to be about 5% to 10% in the work they have done.

5.8 Conclusion

The type of technology and performance evaluation suggested in this and the other chapters makes possible the design of ever more capable networks and grids. These can be used as the foundation of complex technological operations and systems in service to humanity.

5.9 Problems

1. What is a divisible load?
2. What is it about divisible load scheduling modeling that makes solutions tractable? To what other types of modeling is it related?
3. What is the difference between sequential and simultaneous distribution?
4. Intuitively, why is simultaneous distribution scalable?
5. What is the difference between staggered start and simultaneous start?
6. What does a front-end processor in a processor allow one to do? What should lead to a faster solution: the inclusion of a front-end processor or its absence? Why?
7. What is speedup? What does it measure?
8. Explain the Gantt chart-like diagrams used in this chapter.
9. Why does speedup increase nonlinearly for nonlinear models? Is one getting something for nothing?
10. Explain the concept of equivalent processors. How does it help in finding overall network performance?
11. Intuitively, why does speedup saturate as the size of a linear daisy chain is increased (up to infinite size chains)? Why does speedup saturate in tree networks as the number of children and/or levels is increased?
12. Why is it sometimes not worth distributing load to a neighboring node in a linear daisy chain?
13. Why is the environment in which divisible load is processed often time varying?
14. Explain the different parts of Figure 5.16.
15. Why is the equivalence between many divisible load scheduling policies and Markov chains surprising? Why, to some extent, is it not surprising?
16. Do all divisible load scheduling policies have a Markov chain analog? Explain.
17. Verify equation (5.21).
18. Use the result of the equation before equations (5.113) as well as equation (5.112) to verify equation (5.113).
19. Use the two equations before equation (5.129) to verify equation (5.129).
20. Use the two equations before equation (5.167) to verify equation (5.167).
21. Verify equation (5.176).
22. Verify equations (5.190) and (5.191).
23. Derive equation (5.196)

24. Consider a linear daisy chain with N processors where load originates at the left—most processor. Store and forward switching and staggered start is used—a node receives all load for itself and its right neighbors before commencing processing.

    (a) Suppose each processor receives the entire measurement load for processing (perhaps each processor processes the load with different algorithms). Find the optimal number of processors that minimizes the finish time. Note: Using too many processors leads to excessive communication delays and using too few processors leads to insufficient parallelism.

    (b) Suppose that the load is divided into fragments of equal size with one fragment assigned to each processor. Find the optimal number of processors that minimizes finish time.

    (c) Suppose for the situation of (b) that there is non-negligible solution reporting time. Starting with the right-most processor, each processor reports back its solution in time T_s. Solutions are reported back in the opposite order in which load is distributed. Thus the left-most processor eventually receives solutions of duration $(N - 1)T_s$. Find a means of calculating the optimal number of processors.

25. Consider a three children node homogeneous single level tree network. Let $z = 2$, $w = 6$ and $T_{cp} = T_{cm} = 1$. Find the optimal load fractions, α_i's, the equivalent (inverse) processing speed, w_{eq} and speedup for scheduling policies with:

    (a) Sequential distribution.

    (b) Simultaneous distribution and staggered start.

    (c) Simultaneous distribution and simultaneous start.

26. Consider a single level tree network of a control (root) processor and n children processors. Sequential distribution and staggered start is used. The control processor distributes load to its children but does no processing of its own. Draw the Gantt chart timing diagram.

    (a) Write the timing equations.

    (b) Solve for the optimal fraction of load to allocate to the ith processor α_i.

    (c) Find the minimum finish time and speedup.

27. Demonstrate for a linear daisy chain network that, for a minimal finish time solution, all processors must stop processing at the same time. There is no solution reporting time. Load originates at the left-most processor. Do the (partial) proof in the context of the two right-most processors. Assume that processor $N - 1$ keeps a fraction α of the data that it has received and transmits the remaining fraction $1 - \alpha$ to the Nth processor. There are two possibilities, either the N-1st or the Nth processor will stop computing first.

28. Consider two processors without front-end processors and the link connecting them. Processor 1 processes load fraction α_1 and processor 2 processes load fraction $(1 - \alpha_1)$. Draw the Gantt chart timing diagram.

(a) Write the timing equations of the system.

(b) Show that, for parallel processing to save time, $w_1 T_{cp} > z T_{cm}$. That is, the link must be faster than computing at processor 1.

29. (a) Show for a single level tree network that the speedup equation for staggered start reduces to that for simultaneous start if link speed goes to infinity. Intuitively, why is this so?

(b) Show for a single level tree network that the speedup of simultaneous start is always larger than that for staggered start.

30. Consider a linear daisy chain network with front-end processors that is infinite in size in both directions where load originates at an interior processor. The load originating processor first distributes load to its left side and then to its right side. Having a front end, the originating processor computes as it distributes load. Each processor receiving load receives it for itself and for the processors beyond it. Staggered start is used for nodes receiving load. Draw a Gantt chart diagram of the originating processor and its two neighbors.

(a) Write the timing equation of the originating processor and its two neighbors. Let β_c be the amount of total load kept by the originating processor, let β_l be the amount of total load sent to the originating processor's left neighbor, and let β_r be the amount of total load sent to the originating processor's right neighbor. Naturally $\beta_c + \beta_l + \beta_r = 1$.

(b) Write an implicit equation for w_{eqs}^∞, the equivalent inverse speed constant of the three processors.

(c) Suggest a numerical solution technique using the result for load distribution from the boundary of a linear daisy chain of an infinite-sized network as in section 5.4.1.

31. Repeat the steps of the previous problem for load origination at an interior processor in an infinite-sized linear daisy chain network *without* front-end processors. That is, everything is the same as in the previous problem except that the load originating processor does not start computing until it has distributing load to its left and right neighbors.

32. Consider an infinite-sized binary tree network as in subsection 5.4.2 except that processors do not have front-end processors. That is, a node first distributes load to its left child, then to its right child, and only then processes its own fraction of load. Staggered start is used for the children. Draw the Gantt chart of the root node and its children.

(a) Write the timing equations.

(b) Use the result of (a) to write an implicit equation for w_{eq}^∞, the equivalent inverse processing speed of the infinite-sized network.

(c) Using (b), find a polynomial equation that can be solved for w_{eq}^∞.

33. Consider a system as in section 5.5 but with time-varying channel speed, rather than with processor speeds. The channel (bus) is shared with the divisible job of interest and other background transmissions in a "processor sharing" service discipline manner. That is, each of n transmissions on the channel receives $1/n$ of the effort.

(a) Write a similar set of equations to those in section 5.5 for this situation.

(b) Outline a solution algorithm similar to that in section 5.5.

34. Write a set of equations and algorithm for a system as in section 5.5 and the previous problem with both time-varying processor and channel speed. Note: Summarizing previous results, only two equations and some explanation are needed.

35. Phrase the scheduling model for sequential distribution in a single level tree (subsection 5.2.1) as a linear programming problem.

36. Draw the Gantt chart for a single level tree of four children nodes with simultaneous distribution and staggered start. After computation is finished the ith child reports back a solution in time $\alpha_i z_i T_{cm-out}$ (i.e., solution reporting time for a processor is proportional to the load fragment size assigned to the processor). The root does no processing; it only does load distribution to the children nodes.

(a) Solve the model for the optimal amount of load to assign to each processor.

(b) Find the equivalent processing speed of the network as well as its speedup.

37. Draw a Gantt chart for a single level tree with m children with simultaneous distribution and simultaneous start. Here the load must be completely received by the root from an out of network node over a link with inverse transmission speed z_0 before it begins to distribute load to its children. But the root commences processing as it begins to receive load.

(a) Solve the model for the optimal amount of load to assign to each processor.

(b) Find the equivalent processing speed of the network as well as its speedup.

38. Consider N source nodes and M sink nodes. Source nodes distribute divisible load to the sink nodes that do the actual processing. There is a link of inverse speed $z_{i,j}$ between each source node i and each sink node j. The jth sink node has inverse processing speed w_j. A load in the amount L_i is distributed by the ith source node. Also $\sum_{i=1}^{N} L_i = L$. Finally, $\alpha_{i,j}$ is the amount of load that sink j receives from source i and α_j is the fraction of load L that sink j will receive from all of the sources ($\sum_{j=1}^{M} \alpha_j = 1$).

(a) Draw a network diagram and Gantt chart of this situation.

(b) Solve for the optimal amount of load to allocate to each sink and the finish (solution) time.

(c) Does this problem have a unique solution? Comment on this.

A

Summation Formulas

The following summation formulas are used in the text:

$$\sum_{n=0}^{\infty} x^n = \frac{1}{1-x} \qquad 0 \le x < 1 \tag{A.1}$$

$$\sum_{n=0}^{N} x^n = \frac{1 - x^{N+1}}{1-x} \tag{A.2}$$

$$\sum_{n=0}^{\infty} n x^n = \frac{x}{(1-x)^2} \qquad 0 \le x < 1 \tag{A.3}$$

$$\sum_{n=0}^{\infty} n^2 x^n = \frac{x(1+x)}{(1-x)^3} \qquad 0 \le x < 1 \tag{A.4}$$

$$\sum_{n=0}^{\infty} \frac{1}{n!} x^n = e^x \tag{A.5}$$

References

1. B. Abernathy and T.G. Robertazzi, "Loading and Spatial Location in Wire and Radio Communication Networks," *IEEE MILCOM '91*, 1991, pp. 391–395.
2. N. Abramson, "The ALOHA System - Another Alternative for Computer Communications," *Proceedings of the Fall Joint Computer Conference*, 1970.
3. N. Abramson, "The Development of the ALOHANET," *IEEE Transactions on Information Theory*, vol. 31, 1985, pp. 119–123.
4. M. Adler, Y. Gong and A.L. Rosenberg, "Optimal Sharing of Bags of Tasks in Heterogeneous Clusters," *Proc. of ACM SPAA '03*, June 2003.
5. H. Ahmadi and W.E. Denzel, "A Survey of Modern High-Performance Switching Techniques," *IEEE Journal on Selected Areas in Communications*, vol. 7, no. 7, Sept. 1989, pp. 1091–1103.
6. R. Agrawal and H.V. Jagadish, "Partitioning Techniques for Large Grained Paralleism," *IEEE Transactions on Computers*, vol. 37, no. 12, Dec. 1988, pp. 1627–1634.
7. J.R. Artalejo, "G-networks: A Versatile Approach for Work Removal in Queueing Networks," *European Journal of Operations Research*, vol. 126, issue 2, Oct. 2000, pp. 233–249.
8. M. Baker, A. Apon, C. Feiner and J. Brown, "Emerging Grid Standards," *Computer*, vol. 38, no. 4, 2005, pp. 43–50.
9. F. Baskett, K.M. Chandy, R.R. Muntz and F. Palacios, "Open, Closed and Mixed Networks of Queues with Different Classes of Customers," *Journal of the ACM*, vol. 22, no. 2, April 1975, pp. 248–260.
10. S. Bataineh and T.G. Robertazzi, "Bus Oriented Load Sharing for a Network of Sensor Driven Processors," *Special Issue on Distributed Sensor Networks of the IEEE Transactions on Systems, Man and Cybernetics*, vol. 21, no.5, Sept. 1991, pp. 1202–1205.
11. S. Bataineh and T.G. Robertazzi, "Performance Limits for Processor Networks with Divisible Jobs," *IEEE Transactions on Aerospace and Electronic Systems*, vol. 33, no. 4, Oct. 1997, pp. 1189–1198.
12. S. Bataineh, T. Hsiung and T.G. Robertazzi, "Closed Form Solutions for Bus and Tree Networks of Processors Load Sharing a Divisible Job,"

IEEE Transactions on Computers, vol. 43, no. 10, Oct. 1994, pp. 1184–1196.

13. O. Beaumont, L. Carter, J. Ferrante, A. Legrand and Y. Robert, "Bandwidth-Centric Allocation of Independent Tasks on Heterogeneous Platforms," *Proceedings of the International Parallel and Distributed Processing Symposium (IPDPS'02)*, June 2002.

14. D. Bertsekas and R. Gallager, *Data Networks*, 2nd ed., Prentice-Hall, 1991.

15. V. Bharadwaj, D. Ghose and V. Mani, "Optimal Sequencing and Arrangement in Distributed Single-Level Tree Networks with Communication Delays," *IEEE Transactions on Parallel and Distributed Systems*, vol. 5, no. 9, pp. Sept. 1994, pp. 968–976.

16. V. Bharadwaj, D. Ghose and V. Mani, "Multi-installment Load Distribution in Tree Networks with Delays," *IEEE Transactions on Aerospace and Electronic Systems*, vol. 31, no. 2, April 1995, pp. 555–567.

17. V. Bharadwaj, X. Li and K.C. Chung, "Design and Analysis of Load Distribution Strategies with Start-up Costs in Scheduling Divisible Loads on Distributed Networks," *Mathematical and Computer Modelling*, Pergamon Press, April 1999.

18. V. Bharadwaj, X. Li and C.C. Ko, "Efficient Partitioning and Scheduling of Computer Vision and Image Processing Data on Bus Networks using Divisible Load Analysis," *Image and Vision Computing*, vol. 18 no. 11, Aug. 2000a, pg. 919.

19. V. Bharadwaj, H.F. Li and T. Radhakrishnan, "Scheduling Divisible Loads in Bus Networks with Arbitrary Processor Release Times," *Computers and Mathematics with Applications*, Pergamon Press, vol. 32, no. 7, 1996a, pp. 57–77.

20. V. Bharadwaj, D. Ghose, V. Mani and T.G. Robertazzi, *Scheduling Divisible Loads in Parallel and Distributed Systems*, IEEE Computer Society Press, Los Alamitos CA, Sept. 1996b, 292 pages.

21. V. Bharadwaj and N. Viswanadham, "Sub-Optimal Solutions Using Integer Approximation Techniques for Scheduling Divisible Loads on Distributed Bus Networks," *IEEE Transactions on Systems, Man, and Cybernetics: Part A*, vol. 30, no. 6, November 2000b, pp. 680–691.

22. V. Bharadwaj, X. Li and K.C. Chung, "On the Influence of Start-up Costs in Scheduling Divisible Loads on Bus Networks," *IEEE Transactions on Parallel and Distributed Systems*, vol. 11, no. 12, pp. Dec. 2000c, pp. 1288–1305.

23. V. Bharadwaj and S. Ranganath, "Theoretical and Experimental Study of Large Size Image Processing Applications using Divisible Load Paradigm on Distributed Bus Networks," *Image and Vision Computing*, Elsevier Publishers, vol. 20, issues 13–14, Dec. 2002, pp. 917–936.

24. V. Bharadwaj, D. Ghose and T. G. Robertazzi, "A New Paradigm for Load Scheduling in Distributed Systems," *Special Issue of Cluster Computing on Divisible Load Scheduling*, Kluwer Academic Publishers, vol. 6, no. 1, Jan. 2003a, pp. 7–18.

25. V. Bharadwaj and G. Barlas, "Scheduling Divisible Loads with Processor Release Times and Finite Size Buffer Capacity Constraints," *Special Issue of Cluster Computing on Divisible Load Scheduling*, Kluwer Academic Publishers, vol. 6, no. 1, Jan. 2003b, pp. 63–74.

26. J. Blazewicz and M. Drozdowski, "Scheduling Divisible Jobs on Hypercubes," *Parallel Computing*, vol. 21, 1995, pp. 1945–1956.

27. J. Blazewicz and M. Drozdowski, "The Performance Limits of a Two-Dimensional Network of Load Sharing Processors," *Foundations of Computing and Decision Sciences*, vol. 21, no. 1, 1996, pp. 3–15.

28. J. Blazewicz and M. Drozdowski, "Distributed Processing of Divisible Jobs with Communication Startup Costs," Discrete Applied Mathematics, vol. 76, issue 1–3, June 13, 1997, pp. 21–41.

29. J. Blazewicz, M. Drozdowski, F. Guinand and D. Trystram, "Scheduling a Divisible Task in a 2-Dimensional Mesh," Discrete Applied Mathematics, May 1999, pg. 35.

30. S.C. Bruell and G. Balbo, *Computational Algorithms for Closed Queueing Networks*, North-Holland, 1980.

31. J.P. Buzen, "Computational Algorithms for Closed Queueing Networks with Exponential Servers," *Communications of the ACM*, vol. 16, no. 9, Sept. 1973, pp. 527–531.

32. X. Chao and M. Pinedo, "On G-Networks: Queues with Positive and Negative Arrivals," *Probability in the Engineering and Information Sciences*, 1993.

33. S. Charcranoon, T.G. Robertazzi and S. Luryi, "Parallel Processor Configuration Design with Processing/Transmission Costs," *IEEE Transactions on Computers*, vol. 49, no. 9, Sept. 2000, pp. 987–991.

34. S. Charcranoon, T.G. Robertazzi and S. Luryi, "Load Sequencing for a Parallel Processing Utility," *Journal of Parallel and Distributed Computing*, vol. 64, 2004, pp. 29–35.

35. Y.C. Cheng and T.G. Robertazzi, "Distributed Computation with Communication Delays," *IEEE Transactions on Aerospace and Electronic Systems*, vol. 24, no. 6, Nov. 1988, pp. 700–712.

36. Y.C. Cheng, and T.G. Robertazzi, "Distributed Computation for a Tree Network with Communication Delays," *IEEE Transactions on Aerospace and Electronic Systems*, vol. 26, no. 3, May 1990a, pp. 511–516.

37. Y.C. Cheng and T.G. Robertazzi, "A New Spatial Point Process for Multihop Radio Network Modeling," *Proceedings of the International Conference on Communications*, ICC '90, 1990b, pp. 1241–1245.

38. P. Diggle, *Statistical Analysis of Spatial Point Patterns*, Academic Press, 1983.

39. E.W. Dijkstra, "Cooperating Sequential Processes," in F. Genuys, editor, Academic Press, 1968.

40. R. Disney, Traffic Processes in Queueing Networks: A Markov Renewal Approach, The Johns Hopkins University Press, Baltimore MD, 1987.

41. M. Drozdowski, *Selected Problems of Scheduling Tasks in Multiprocessor Computer Systems*, Politechnika Poznanska, Book No. 321, Poznan, Poland, 1997.

42. M. Drozdowski and W. Glazek, "Scheduling Divisible Loads in a Three Dimensional Mesh of Processors," *Parallel Computing*, vol. 25, 1999, pp. 381–404.

43. M. Drozdowski and P. Wolniewicz, "Experiments with Scheduling Divisible Tasks in Clusters of Workstations," in A. Bode, T. Ludwig, W. Karl and R. Wismüler, editors, *EURO-Par 2000*, Lecture Notes in Computer Science 1900, Springer-Verlag, 2000, pp. 311–319.

44. M. Drozdowski and P. Wolniewicz, "Divisible Load Scheduling in Systems with Limited Memory," *Special Issue of Cluster Computing on Divisible Load Scheduling*, Kluwer Academic Publishers, vol. 6, no. 1, Jan. 2003, pp. 19–30.

45. P.-F. Dutot, "Divisible Load on Heterogeneous Linear Array," Proceedings of the International Parallel and Distributed Processing Symposium (IPDPS'03), Nice, France, April 2003.

46. C. Eklund, R.B. Marks, K.L. Stanwood and S. Wang, "IEEE Standard 802.16: A Technical Overview of the WirelessMAN Air Interface for Broadband Wireless Access," *IEEE Communications Magazine*, June 2002, pp. 98–107.

47. I. Foster and C. Kesselman, The Grid 2: Blueprint for a New Computing Infrastructure, Morgan-Kaufman, 2003.

48. M.A. Franklin, "A VLSI Performance Comparison of Banyan and Crossbar Communication Networks," *IEEE Transactions on Computers*, vol. C-30, no. 4, April 1981, pp. 283–290.

49. M. W. Garrett and S.-Q. Li, "A Study of Slot Reuse in Dual Bus Multiple Access Networks," *Proceedings of INFOCOM '90*, San Francisco, CA, June 1990.

50. T. Garritano, "Globus: An Infrastructure for Resource Sharing," *Clusterworld*, vol. 1, no. 1, pp. 30–31, 50.

51. E. Gelenbe, "Product Form Networks with Negative and Positive Customers," *Journal of Applied Probability*, vol. 28, 1991a, pp. 656–663.

52. E. Gelenbe, P. Glynn and K. Sigman, "Queues with Negative Arrivals," *Journal of Applied Probability*, vol. 28, 1991b, pp. 245–250.

53. D. Ghose, "A Feedback Strategy for Load Allocation in Workstation Clusters with Unknown Network Resource Capabilities using the DLT Paradigm," *Proceedings of the International Conference on Parallel and Distributed Processing Techniques and Applications (PDPTA'02)*, Las Vegas, Nevada, vol. 1, June 2002, pp. 425–428.

54. D. Ghose and H.J. Kim, "Load Partitioning and Trade-Off Study for Large Matrix Vector Computations in Multicast Bus Networks with Communication Delays," *Journal of Parallel and Distributed Computing*, vol. 54, 1998.

55. D. Ghose and V. Mani, "Distributed Computation with Communication Delays: Asymptotic Performance Analysis," *Journal of Parallel and Distributed Computing*, vol. 23, 1994, pp. 293–305.

56. D. Ghose and T.G. Robertazzi, editors, *Special Issue on Divisible Load Scheduling, Cluster Computing*, vol. 6, 2003.

57. W. Glazek, "A Multistage Load Distribution Strategy for Three Dimensional Meshes," *Special Issue of Cluster Computing on Divisible Load Scheduling*, Kluwer Academic Publishers, vol. 6, no. 1, Jan. 2003, pp. 31–40.

58. L. Goldberg, "802.11 Wireless LANs: A Blueprint for the Future?," *Electronic Design*, Aug. 4, 1997, pp. 44–52.

59. D. Goodman and R. Yates, *Probability and Stochastic Processes*, 2nd ed., Wiley, 2004.

60. W.J. Gordon and G.F. Newell, "Closed Queueing Systems with Exponential Servers," *Operations Research*, vol. 15, 1967, pp. 254–265.

61. D. Gross and C.M. Harris, *Fundamentals of Queueing Theory*, Wiley, 1974, 1985.

62. J.C. Haartsen and S. Mattisson, "Bluetooth - A New Low-Power Radio Interface Providing Short-Range Connectivity," *Proceedings of the IEEE*, vol. 88, no. 10, Oct. 2000, pp. 1651–1661.

63. J.L. Hammond and P.J.P. O'Reilly, *Performance Analysis of Local Computer Networks*, Addison-Wesley, 1986.

64. U. Herzog, L. Woo and K.M. Chandy, "Solution of Queueing Problems by a Recursive Technique," *IBM Journal of Research and Development*, May 1975, pp. 295–300.

65. F.S. Hillier and G.J. Lieberman, Introduction to Operations Research, 8th edition, McGraw-Hill, 2005.

66. G.J. Holzmann, "The Model Checker Spin," *IEEE Transactions on Software Engineering*, vol. 23, no. 5, May 1997, pp. 279–295.

67. G.J. Holzmann, *The SPIN Model Checker: Primer and Reference Manual*, Addison-Wesly, 2004.

68. J.Y. Hui and E. Arthurs, "A Broadband Packet Switch for Integrated Transport," *IEEE Journal on Selected Areas in Communications*, vol. SAC-5, no. 8, Oct. 1987, pp. 1264–1273.

69. J.T. Hung, *Scalable Scheduling in Parallel, Distributed and Grid Systems*, Ph.D Thesis, Dept. of Electrical and Computer Engineering, Stony Brook University, Stony Brook, NY, Aug. 2003b.

70. J.T. Hung and T.G. Robertazzi, "Distributed Scheduling of Nonlinear Computational Loads," *Proceedings of the 2003 Conference on Information Sciences and Systems*, The Johns Hopkins University, Baltimore, MD, March 2003a.

71. J.T. Hung and T.G. Robertazzi, "Scalable Scheduling for Clusters and Grids using Cut Through Switching," *International Journal of Computers and their Applications*, ACTA Press, vol. 26, no. 3, 2004a, pp. 147–156.

72. J.T. Hung and T.G. Robertazzi, "Divisible Load Cut Through Switching in Sequential Tree Networks," *IEEE Transactions on Aerospace and Electronic Systems*, vol. 40, no. 3, July 2004b, pp. 968–982.

73. J.T. Hung, H.J. Kim and T.G. Robertazzi, "Scalable Scheduling in Parallel Processors," *Proceedings of the 2002 Conference on Information Sciences and Systems*, Princeton University, Princeton, NJ, March 2002.

74. J.R. Jackson, "Networks of Waiting Lines," *Operations Research*, vol. 5, 1957, pp. 518–521.

75. J.M. Kahn, R.H. Katz and K.S.J. Pister, "Emerging Challenges: Mobile Networking for 'Smart Dust'," *Journal of Communications Networks*, vol. 2, no. 3, Sept. 2000.

76. S. Kapp, "802.11: Leaving the Wire Behind," *IEEE Internet Computing*, vol. 6, no. 1, Jan.–Feb. 2002, pp. 82–85.

77. M.J. Karol, M.G. Hluchyj and S.P. Morgan, "Input vs. Output Queueing on a Space Division Packet Switch," *IEEE Transactions on Communications*, vol. COM-35, no. 12, Dec. 1987, pp. 1345–1356.

78. J.S. Kaufman, "Blocking in a Shared Resource Environment," *IEEE Transactions on Communications*, vol. COM-29, no. 10, Oct. 1981, pp. 1474–1481.

79. D.G. Kendall, "Some Problems in the Theory of Queues," *Journal of the Royal Statistical Society*, series B, vol. 13, no. 2, 1951, pp. 151–185.

80. P. Kermani and L. Kleinrock, "Virtual Cut-Through: A New Computer Communications Switching Technique," *Computer Networks*, vol. 3, 1979, pp. 267–286.

81. A.Y. Khinchin, "Mathematisches uber die Erwatung vor einem offentlichen Schalter," in English "Mathematical Theory of Stationary Queues," *Matem. Sbornik*, vol. 39, 1932, pp. 73–84.

82. H.J. Kim, "A Novel Optimal Load Distribution Algorithm for Divisible Loads," *Special Issue of Cluster Computing on Divisible Load Scheduling*, Kluwer Academic Publishers, vol. 6, no. 1, Jan. 2003, pp. 41–46.

83. H.J. Kim, G.-I. Jee and J.G. Lee, "Optimal Load Distribution for Tree Network Processors," *IEEE Transactions on Aerospace and Electronic Systems*, vol. 32, no. 2, April 1996, pp. 607–612.

84. S.-H. Kim and T.G. Robertazzi, "Spatial Network Traffic Intensity," *Proceedings of the 2000 Conference on Information Sciences and Systems*, Princeton University, Princeton, NJ, 2000, pp. TP2–24.

85. L. Kleinrock, *Queueing Systems*, vol. I and II, Wiley, 1975.

86. K. Ko, *Scheduling Data Intensive Parallel Processing in Distributed and Networked Environments*, Ph.D Thesis, Dept. of Electrical and Computer Engineering, Stony Brook University, Stony Brook, NY, Aug. 2000a.

87. K. Ko and T.G. Robertazzi, "Record Search Time Evaluation," *Proceedings of the Conference on Information Sciences and Systems*, Princeton University, Princeton, NJ, March 2000b.

88. K. Ko and T.G. Robertazzi, "Scheduling in an Environment of Multiple Job Submission," *Proceedings of the 2002 Conference on Information Sciences and Systems*, Princeton University, Princeton, NJ, March 2002.

89. H. Kobayashi, *Modeling and Analysis: An Introduction to System Performance Evaluation*, Addison-Wesley, 1978.

90. C.S. Kong, V. Bharadwaj and D. Ghose, "Large Matrix-Vector Products on Distributed Bus Networks with Communication Delays using the Divisible Load Paradigm: Performance and Simulation," *Computers and Mathematics in Simulation*, Elsevier Press, vol. 58, 2001, pp. 71–92.

91. B. Kreaseck, L. Carter, H. Casanova and J. Ferrante, "Autonomous Protocols for Bandwidth-Centric Scheduling of Independent-Task Applications," *Proceedings of the International Parallel and Distributed Processing Symposium (IPDPS'03)*, Nice, France, April 2003.

92. R.O. LaMaire, A. Krishnan, P. Bhagwat and J. Panian, "Wireless LANs and Mobile Networking: Standards and Future Directions," *IEEE Communications Magazine*, vol. 34, no. 8, Aug. 1996, pp. 86–94.

93. C.E. Leiserson, "Fat-Trees: Universal Networks for Hardware-Efficient Supercomputing," *IEEE Transactions on Computers*, vol. C-34, no. 10, 1985, pp. 892–901.

94. X. Li, V. Bharadwaj and C.C. Ko, "Optimal Divisible Task Scheduling on Single-Level Tree Networks with Finite Size Buffers," *IEEE Transactions on Aerospace and Electronic Systems*, vol. 36, no. 4, Oct. 2000, pp. 1298–1308.

95. M. Littlewood and I.D. Gallagher, "Evolution Toward an ATD Multiservice Network," *British Telecom Technology Journal*, vol. 5, no. 2, April 1987.

96. V. Mani and D. Ghose, "Distributed Computation in a Linear Network: Closed-Form Solutions and Computational Techniques," *IEEE Transactions on Aerospace and Electronic Systems*, vol. 30, no. 2, April 1994.

97. M.A. Marsan, G. Balbo and G. Conte, *Performance Models of Multiprocessor Systems*, The MIT Press, 1986.

98. M.A. Marsan, G. Balbo, G. Conte and F. Gregoretti, "Modeling Bus Contention and Memory Interference in a Multiprocessor System," *IEEE Transactions on Computers*, vol. C-32, no. 1, 1983, pp. 60–72.

99. M. Mauve, H. Hastenstein and A. Widmer, "A Survey on Position-Based Routing in Mobile Ad Hoc Networks," *IEEE Network*, vol. 15, no. 3, Nov./Dec. 2001, pp. 30–39.

100. H. Michiel and K. Laevens, "Teletraffic Engineering in a Broad-Band Era," Proceedings of the IEEE, vol. 85, no. 12, Dec. 1997, pp. 2007–2033.

101. L.E. Miller, "Distributional Properties of Inhibited Random Positions of Mobile Radio Terminals," *Proceedings of the 2002 Conference on Information Sciences and Systems*, Princeton University, Princeton, NJ, March 2002.

102. M. Moges and T.G. Robertazzi, "Optimal Divisible Load Scheduling and Markov Chain Models," *Proceedings of the 2003 Conference on Information Sciences and Systems*, The Johns Hopkins University, Baltimore, MD, March 2003.

103. C.S.R. Murthy and B.S. Manoj, *Ad Hoc Wireless Networks: Architectures and Protocols*, Prentice-Hall, 2004.

104. Y. Oie, M. Murata, K. Kubota and H. Miyahara, "Effect of Speedup in Nonblocking Packet Switches," *Proceedings of IEEE International Conference on Communications*, 1989, pp. 410–414.

105. R.O. Onvural, *Asynchronous Transfer Mode Networks: Performance Issues*, 2nd ed., Artech House, 1995.

106. A. Papoulis and S.U. Pillai, *Probability, Random Variables and Stochastic Processes*, McGraw-Hill, 2002.

107. J.H. Patel, "Performance of Processor-Memory Interconnection for Multiprocessors, *IEEE Transactions on Computers*, vol. C-30, no. 10, Oct. 1981, pp. 771–780.

108. C.E. Perkins, *Ad Hoc Networking*, Addison-Wesley, 2000.

109. W.W. Peterson and D.T. Brown, "Cyclic Codes for Error Detection," *Proceedings of the IRE (Institute of Radio Engineers)*, 1961, pp. 228–235.

110. C.A. Petri, *Kommunikation mit Automaten*, Ph.D Thesis, University of Bonn, Germany, 1962.

111. D.A.L. Piriyakumar and C.S.R. Murthy, "Distributed Computation for a Hypercube Network of Sensor-Driven Processors with Communication Delays Including Setup Time," *IEEE Transactions on Systems, Man and Cybernetics-Part A: Systems and Humans*, vol. 28, no. 2, March 1998, pp. 245–251.

112. F. Pollaczek, "Uber eine Aufgabe dev Wahrscheinlichkeitstheorie," *Math. Zeitschrift*, vol. 32, 1930, pp. 64–100, 729–750.

113. J.M. Rabaey, M.J. Ammer, J.L. da Silva Jr., et. al., "PicoRadio Supports Ad Hoc Ultra-Low Power Wireless Networking," *Computer*, vol. 33, no. 7, July 2000, pp. 42–48.

114. M. Reiser and H. Kobayashi, "Recursive Algorithms for General Queueing Networks with Exponential Servers," *IBM Research Report RC 4254*, Yorktown Heights, NY, 1973.

115. M. Reiser and S.S. Lavenberg, "Mean-Value Analysis of Closed Multichain Queueing Networks," *Journal of the ACM*, vol. 27, no. 2, April 1980, pp. 313–322.

116. T.G. Robertazzi, *Performance Evaluation of High Speed Switching Fabrics and Networking: ATM, Broadband ISDN and MAN Technology*, IEEE Press, 1993a (now distributed by Wiley).

117. T.G. Robertazzi, "Processor Equivalence for a Linear Daisy Chain of Load Sharing Processors," *IEEE Transactions on Aerospace and Electronic Systems*, vol. 29, no. 4, Oct. 1993b, pp. 1216–1221.

118. T.G. Robertazzi, Planning Telecommunication Networks, Wiley and IEEE Press, 1999.

119. T.G. Robertazzi, *Computer Networks and Systems: Queueing Theory and Performance Evaluation*, 3rd ed., Springer-Verlag, 2000.

120. T.G. Robertazzi, "Ten Reasons to Use Divisible Load Theory," *Computer*, vol. 36, no. 5, May 2003, pp. 63–68.

121. L. Roberts, "Extensions of Packet Communication Technology to a Hand Held Personal Terminal," *Proceedings of the Spring Joint Computer Conference*, AFIPS, 1972, pp. 295–298.

122. M. A. Rodrigues, "Erasure Node: Performance Improvements for the IEEE 802.6 MAN," *Proceedings of IEEE INFOCOM '90*, San Francisco, CA, June 1990, pp. 636–643.

123. R. Rom and M. Sidi, *Multiple Access Protocols: Performance and Analysis*, Springer-Verlag, 1990.

124. K.W. Ross, *Multiservice Loss Models for Broadband Telecommunication Networks*, Springer Verlag, 1997.

125. J.P. Ryan, "WDM: North American Deployment Trends," *IEEE Communications Magazine*, vol. 36, no. 2, Feb. 1998, pp. 40–44.

126. T.N. Saadawi, M.H. Ammar and A. El Hakeem, *Fundamentals of Telecommunication Networks*, Wiley, 1994.

127. J.M. Schopf and B. Nitzberg, "Grids: The Top Ten Questions," *Scientific Programming*, IOS Press, vol. 10, no. 2, 2002, pp. 103–111.

128. M. Schwartz, *Telecommunication Networks: Protocols, Modeling and Analysis*, Addison-Wesley, 1987.

129. L. Schwiebert, S.K.S. Gupta and J. Weinmann, "Research Challenges in Wireless Networks of Biomedical Sensors," *ACM Sigmobile*, 2001, pp. 151–165.

130. R.C. Shah and J.M. Rabaey, "Energy Aware Routing for Low Energy Ad Hoc Sensor Networks," *Proceedings of the 3rd IEEE Wireless, Communications and Networking Conference*, 2002, pp. 350–355.

131. C.E. Siller, editor, *SONET/SDH: A Sourcebook of Synchronous Networking*, Wiley-IEEE Press, 1996.

132. S. Siwamogsatham, "10 Gigabit Ethernet," *www.cse.wustl.edu/j̃ain/cis788-99/ftp/10gbe/index.html*, 1999.

133. J. Sohn and T.G. Robertazzi, "Optimal Load Sharing for a Divisible Job on a Bus Network," *IEEE Transactions on Aerospace and Electronic Systems*, vol. 32, no. 1, Jan. 1996, pp. 34–40.

134. J. Sohn and T.G. Robertazzi, "Optimal Time Varying Load Sharing for Divisible Loads," *IEEE Transactions on Aerospace and Electronic Systems*, vol. 34, no. 3, July 1998b, pp. 907–924.

135. J. Sohn, T.G. Robertazzi and S. Luryi, "Optimizing Computing Costs using Divisible Load Analysis," *IEEE Transactions on Parallel and Distributed Systems*, vol. 9, March 1998a, pp. 225–234. Also related: T.G. Robertazzi, S. Luryi and J. Sohn, Load Sharing Controller for Optimizing Monetary Cost, US Patent 5,889,989, March 30, 1999. T.G. Robertazzi, S. Luryi and S. Charcranoon, Load Sharing Controller for Optimizing Resource Utilization Cost, US Patent 6,370,560, April 9, 2002.

136. W. Stallings, *High-Speed Networks and Internets: Performance and Quality of Service*, Prentice-Hall, 2002.

137. T.H. Szymanksi, "A VLSI Comparison between Crossbar and Switch-Recursive Banyans Interconnection Networks," Proceedings of the International Conference on Parallel Processing, Aug. 1986, pp. 192–199.

138. A.S. Tanenbaum, *Computer Networks*, 3rd ed., Prentice-Hall, 1996.

139. A.S. Tanenbaum, *Computer Networks*, 4th ed., Prentice-Hall, 2003.

140. C.-K. Toh, *Ad Hoc Mobile Wireless Networks: Protocols and Systems*, Prentice-Hall, 2002.

141. S.J. Vaughan-Nichols, "Will 10-Gigabit Ethernet Have a Bright Future?," *Computer*, June 2002, pp. 22–24.

142. I.Y. Wang and T.G. Robertazzi, "Recursive Computation of Steady State Probabilities of Non-Product Form Queueing Networks Associated with Computer Network Models," *IEEE Transactions on Communications*, vol. 38, no. 1, Jan. 1990, pp. 115–117.

143. J. Williams, "The 802.11 b Security Problem – Part I," IEEE ITPro (Information Technology Professional), Nov/Dec 2001, pp. 91–96.

144. M.E. Woodward, Communication and Computer Networks: Modelling with Discrete-Time Queues, Wiley and IEEE Computer Society Press, 1993.

145. C.-I. Wu and T.-Y. Feng, *Tutorial: Interconnection Networks for Parallel and Distributed Processing*, IEEE Computer Society Press, 1984.

146. L. Xiaolin, V. Bharadwaj and C.C. Ko, "Experimental Study on Processing Divisible Loads for Large Size Image Processing Applications using PVM Clusters," International Journal of Computers and Applications, ACTA Press, July 2001.

147. Y.-S. Yeh, M.G. Hluchyj and A.S. Acampora, "The Knockout Switch: A Simple, Modular Architecture for High Performance Packet Switching," *Journal on Selected Areas in Communication*, vol. SAC-5, no. 8, Oct. 1987, pp. 1274–1283.

148. H. Yoon, K.Y. Lee and M.T. Liu, "Performance Analysis of Multi-buffered Packet-Switching Networks in Multiprocessor Systems," *IEEE Transactions on Computers*, vol. 39, no. 3, March 1990, pp. 319–327.

149. M.C. Yuang, "Survey of Protocol Verification Techniques Based on Finite State Machine Models," *Proceedings of Computer Networking Symposium*, Washington, D.C., 1988, pp. 164–172.

150. J. Zheng and M.J. Lee, "Will 802.15.4 Make Ubiquitous Networking A Reality? A Discussion on a Potential Low Power, Low Bit Rate Standard," *IEEE Communications Magazine*, June 2004, pp. 140–146.

Index

10 Gbps Ethernet, 21
10 Mbps Ethernet, 16
802.11a, 24
802.11b, 24
802.11g, 24
802.15.3a, 27
802.15.4, 27
802.16, 28
802.6, 69

ad hoc network, 7, 173
Aloha packet radio, 62
analytical solutions, 152
AODV, 7
ATM, 30
Automatic Repeat Request, 178
available bit rate, 34

Banyan networks, 172, 173
Bernoulli process, 46, 52
binary tree, 72
binomial distribution, 55
blocking probability, 126
Bluetooth, 26

cellular telephone, 5
circuit switching, 11
closed Markovian queueing network,
    132
clusters, 170
coaxial cable, 2
collision, 58
concurrency, 149
confidence intervals, 154

constraints, 256
convolution algorithm, 135
crossbar, 87
CSMA/CD, 57
cyclic redundancy codes, 184

daisy chains, 236
data parallel loads, 193
delta network, 170
departure instants, 122
differential equations, 102
Dijkstra algorithm, 164
discrete event simulation, 153
divisible load scheduling, 193
DQDB, 69

equivalent element, 221
equivalent processors, 197
Erlang B, 117
Erlang C, 119
error detecting codes, 184
Ethernet, 15, 57
Ethernet design equation, 62

Fast Ethernet, 18
fiber optic, 3
flooding, 169
Ford Fulkerson algorithm, 166
frequency hopping, 10
front-end subprocessors, 200

Gantt chart, 201
Generalized stochastic Petri networks,
    148

Geom/Geom/1, 108
geometric distribution, 53
geostationary satellites, 4
Gigabit Ethernet, 20
global balance, 102
global balance equation, 129
Global Grid Forum, 42
grid, 41

Hamming error correcting code, 181
hierarchical routing, 169

IEEE 802.11, 22
IEEE 802.15, 26
infinite-sized networks, 196
inhibitor arcs, 150
Inter Packet Gap, 21
interconnection network, 87
interconnections topologies, 195
interior point method, 255
Iridium, 5
ISM, 23

knockout switch, 84

layered architecture, 13
linear programming, 255
linear topology, 69
local balance, 104
local balance equation, 129
low earth orbit satellites, 4

M/G/1, 122
M/M/1, 100
M/M/1/N, 113
Manchester encoding, 17
Markovian Petri networks, 148
mean service rate, 100
mean value analysis, 135
meta-computing, 198
microwave radio, 4
moment generating functions, 153
multiple access, 57
multiplexing, 9

negative customers, 141
networks of queues, 128
NNI, 32
non-product form models, 144
normalization constants, 122

numerical solution, 152

objective function, 256
OGSA, 42
open queueing network, 128

packet switch, 45
packet switching, 11
parity, 180
parity codes, 181
Pascal distribution, 56
path server, 168
performance evaluation, 46
performance measure, 79
performance measures, 126
Petri networks, 148
piconet, 27
places, 148
Poisson distribution, 50
Poisson process, 49
Pollaczek–Khinchin mean value
    formula, 124
positive customer, 141
priority classes, 144
proactive algorithms, 7
probability, 92
probability flux, 101
product form equations, 132
propagation delay, 61
protection fibers, 39
protocol verification, 174
protocols, 13, 174
pseudo-random numbers, 153

QoS, 34
quality-of-service, 34
queueing theory, 99

random routing, 138
random walks, 102
reachability diagram, 176
reactive algorithms, 7
resource sharing, 149
ring topologies, 39
routing, 161
routing table, 165

satellites, 4
scatternet, 27
self-routing, 170
serializability, 149

service fiber, 39
shared memory switch, 35
simplex algorithm, 255
simulation, 153
SONET, 36
source routing, 168
space division switches, 36
spanning tree, 165
spatial distribution, 75
speedup, 204
SPIN model checker, 178
state transition diagram, 49
STM, 30
switching elements, 79
synchronization, 149
Synchronous Optical Network, 36

table driven routing, 167
teletraffic modeling, 69
time division multiplexing, 9
tokens, 148

torus, 150
transient models, 152
transition, 148
transitions, 101
tree topologies, 71
tree topology, 245
twisted pair, 3

UNI, 32
Unspecified bit rate, 34

virtual tributaries, 37
VLSI, 78

wavelength division multiplexing, 40
WDM, 40
web server, 54
window of vulnerability, 64
wireless sensor networks, 8
wireless technology, 22

Zigbee, 27

Information Technology: Transmission, Processing, and Storage

(continued from page ii)

Nonuniform Sampling: Theory and Practice
Edited by Farokh Marvasti

Simulation of Communication Systems, Second Edition
Michael C. Jeruchim, Phillip Balaban and K. Sam Shanmugan

Principles of Digital Transmission
Sergio Benedetto, Ezio Biglieri

.